Apples

for the

Twenty-First Century

Part III

Part IV

Part V

Part VI

Part VII

Appendices

✳ ✳

Apples

for the

Twenty-First Century

By
Warren Manhart

✳ ✳

Published by the North American Tree Company, 1995

First published in 1995 in Portland, Oregon, by North American Tree Company, D.B.A.
Portland Nursery, 5050 S.E. Stark Street, Portland, Oregon 97215.

Library of Congress Cataloging-in-Publication Data

Manhart, Warren
     Apples for the 21st century / Warren Manhart
     ISBN 0-9648417-0-3

Manufactured in the United States of America

First printing, 1995

This book is dedicated to my wife, Joyce. She put up with an incredible number of apples appearing in September, October and November, finding wonderful uses for them. Then, when it came time to copy my scribbling, her expertise with a word processor became invaluable.

Table of Contents

Part I

Part II

Foreword

Apple growers—amateurs, hobbyists, and semi-professionals—need objective information to help them sort through the many cultivars (varieties) available today.

Regardless of the size of the intended orchard, the planting of any apple tree represents an investment of money, effort and growing space for many years into the future. Commercial orchardists have experience to help them determine what the market will be, while those with less experience rely more on the nurseries they use. Most nurseries stock primarily the cultivars that they think their customers will purchase. The publishers of this book feel it is crucial to the retail buyer, then, that those retail nurseries have objective information available, so that they make the right selections for now and well into the next century.

Warren Manhart has had extensive experience with both the areas east of the Rocky Mountains and the apple regions of the Pacific Northwest. He has worked with fruit in the cold regions of Minnesota as well as the beautiful Willamette Valley of Oregon.

Mr. Manhart knows agriculture first-hand from his early years on the farm in Minnesota. He also did undergraduate work in botany at St. Olaf College in Northfield, Minnesota. Later, he pursued multiple courses in agronomy, horticulture and plant pathology at the University of Minnesota.

Following service in World War II as a Marine Corps pilot, his vocation was in marketing. More than 30 years ago he moved his family west to Oregon, and since that time he has been involved in many things that have fortified his position as an authority on apple varieties. He has taught apple-related courses at three community colleges. Over the past several decades he has planted, harvested, evaluated and replanted many apple cultivars, both old and new.

Warren Manhart has served as president of the Home Orchard Society in the Portland area and as vice president of the Pacific Northwest Fruit Testers Association. He has maintained membership in the International Dwarf Fruit Tree Association as well as the North American Fruit Explorers organization. In 1991 he was awarded the Willamette Valley Fruit Growers Association "Man of the Year" award as recognition of his many years of service to fruit growers in his area. His peers regard him as one of the best apple cultivar identification experts.

We are happy to cooperate in the production of this book because we feel Mr. Manhart has evaluated apple varieties clearly and accurately. Many persons interested in planting apple trees, in whatever amount, can profit from the information included here. Retail nurserymen will find a wealth of valuable information in this book. We recommend that they use it as a reference in their nurseries as well as making it available to their customers.

North American Tree Company
D.B.A. Portland Nursery
5050 S.E. Stark Street
Portland, Oregon 97215

Acknowledgments

All color photographs were taken by the author except five: Wickson Crab was taken by Ram Fishman of Greenmantle Nursery, and Ginger Gold™ was supplied by the patent holder, Adams County Nursery of Pennsylvania. Carousel Apple, supplied by Willow Drive Nursery. Remaining 'Hawkeye Delicious' tree trunk supplied by Robert Denny of North American Tree Co. Pink Lady Apple picture supplied by E. W. Brandt and Sons. The drawings were done by Mark Manhart. Editing by Carolyn Manhart. Final editing and formating was done by Catharine Hess.

Part I

Introduction

The selection of apple cultivars listed in this book is based on testing many of the best cultivars grown worldwide. Almost without exception, all have been grown, studied and photographed in my own orchard over a 30-year period.

This information resulting from my testing will be useful to growers ranging from novices interested in growing apples for home consumption to the small direct-market commercial grower. Both groups share a need for a reliable guide to better apples. Since about one-half of the fruit described is seldom available in United States retail markets, this book can be your introduction to different or perhaps better apples.

The larger commercial orchardists select only a few cultivars that, they hope, are easy to grow, provide large yields, are annual bearers, and ship and store well. They usually underestimate the consumer's desire for good-tasting apples. Let us also put some blame for inferior fruit on the system of middlemen such as packer-sales outlets, brokers and large store buyers, however necessary they seem to many growers who just want to grow and do no marketing. In my opinion, that system needs some decentralization. If we had more "niche" cultivars available locally, we would not have to accept some of the early-picked, stored-for-9-months, starchy disasters that are yearly thrust down the throats of unwilling or unwitting millions. Well, times are changing with the surfacing of great new and old cultivars, some from the southern hemisphere. One wonders why these middlemen have not understood why our country has such low apple consumption compared to Europe. The Europeans have long informed us that more cultivars with good taste would increase consumption considerably. I agree.

The apple produces a remarkably diverse variety of taste, color, storage characteristics and use—fresh dessert, cooking, snacking, cider and drying—yet consumers are offered a small percentage of such variety, except by the direct-market orchardists. Direct marketers know what they can grow with quality results, and what consumers in their area will buy. Taste testing is the quickest and most conclusive way of introducing a new cultivar.

From 1950 to 1980, per-capita apple consumption in North America appeared to be at an all-time low, or about one-third to one-half of that in the northern European countries of Belgium, The Netherlands, Switzerland and northern Germany. Despite considerable advertising and what seems to be a permanent change in the American diet to consume more fruits and vegetables, apple use increased only about two pounds per person from 1980 to 1990. On average, North Americans eat about 20 pounds per person a year for fresh dessert (40 to 70 apples). Approximately another 21 pounds is consumed per capita for pie, sauce, juice, baby food and drying. This totals only one 40-pound box per year (one bushel). This low consumption cannot be attributed to the consumption of other fruits such as bananas, grapes, citrus, tropical fruits, etc., as the Europeans consume as much or more of those than we do.

A few years ago the New Zealanders demonstrated this example from the Europeans by exporting three cultivars on a test basis to the west coast from Vancouver, B.C., to San Diego, California. These were cultivars few North American growers had ever heard of, and had been grown only by long-time testers like myself. Granny Smith was first, followed by Gala and then, 9 years ago, Braeburn. They sold out for 99 cents to $1.89 per pound with no promotion. With great perception the New Zealanders saw a large hole in the U.S. apple market and are now partially filling it, reaping good profits and expanding their production. The fact that they have opposite seasons from North America no doubt helps. Gala arrives on the west coast in March, April and May. Braeburn arrives about late May on through the summer. Granny Smith starts just after Braeburn. These fresh apples compete very well with our cold storage Red Delicious and Golden Delicious, which get a little tired by July. We are now in a world market, with southern hemisphere growers important contributors. South America may soon enter this market in apples. Chile is already a big shipper to us with other fruits, and Argentina may ship us apples in a few years.

Thirty years ago I rebelled against the U.S. grower-packer system with its emphasis on three cultivars and early picking for long storage. I wanted better apples, so I have tested a large number of cultivars, and tried to stay 5 to 10 years ahead of growers' needs. Mine is a test orchard with the best apple cultivars I can find worldwide and grow legally after post-quarantine has made virus-free grafting wood, or grafted trees, available.

I have provided the descriptions and color illustrations to assist home orchardists and direct-market commercial growers in learning which cultivars may do well in their climate. They should also help in identification. These descriptions are based on my own trees, and they may not quite agree with results in other climates. The decisions of which rootstock to use, trellising, pollination, etc., require study and logical decisions about what to grow. These apple cultivars remind me of the needs of children; children all need certain basics, but they are all different and have differing needs beyond those basics.

Most old apple books have included myriads of cultivars, many of which were grown in small areas, sometimes only within one county of one state. In my opinion, it is now unwise and seems counter-productive to describe hundreds of obscure cultivars that, although edible, are often unobtainable and perhaps undesirable. Such descriptions of huge numbers could be very confusing, leading to experimentation and wasted years (even if you could get the trees.) Some apples are primarily fresh dessert, some are great keepers, some are for processing, and a very few, such as Braeburn, Melrose, Spitzenberg, or Newtown, fit all these categories. Only 50 are given complete descriptions. Some for the far north and far south are grown in such small quantities as not to warrant the many more years of research needed.

Dr. Melvin Westwood, retired Oregon State University scientist and author of the fine book *Temperate Zone Pomology*, made the wry observation, "It isn't always what we know that affects our results so much, as what we know that ain't so!" My book should help keep the reader from acting on ideas and procedures that "ain't so." The different climates, soils, and orchard management approaches around our large nation may produce results differing somewhat from mine.

Due to the large number of variables involved such as soil, fertilizer, yearly variation of climate, cultivars, rootstock, microclimates and orchard management, the reader cannot be guaranteed a superb crop every year. A book such as this would have saved me at least 14 years of experimenting with obscure, old, low-quality cultivars and rootstocks, then called semi-dwarfs, that often gave huge trees 12 to 20 feet in height. I refuse to grow apples anymore on trees so large that one risks life and limb taking care of them, and I believe most readers will share these views. We don't have to have such large trees to get superior fruit. On the contrary, the dwarfing, efficient rootstocks frequently give larger and better apples; e.g., the M.9 root.

I have especially had in mind readers over 50 years of age who do not have as much time for experimentation as the younger reader. With apple trees, mistakes don't usually surface for a few years, and then it may take 4 to 6 more years to correct them with re-planting. In the meantime, the grower could become discouraged. When a gardener makes mistakes with annuals such as carrots, squash or beans, corrections can be made the following spring. Not so with an apple tree.

Flavor

Even with the most precocious (early bearing) of rootstock and cultivar combinations, it takes 2 to 4 years after planting a 1-year-old whip before the grower can assess the apple for taste. Flavor is seldom optimum in the first 2 years of fruiting because excess nitrogen is usually used to bring the tree quickly to a height of 6 to 8 feet. The root structure on these young dwarfing trees is usually not adequate to supply proper amounts of calcium and other minerals to the fruit. This is why apples on young trees are so susceptible to bitterpit, a calcium shortage in the apple. I try not to judge an apple's flavor until the tree has fruited at least three times. With Fuji it may take 5 fruiting years, as its young trees produce quite differently than 8- to 10-year-old trees. No one knows the reason for sure why Fuji is so tardy at giving us its best; however, after 12 years of testing one strain of it, I believe the fruit to be very nitrogen sensitive.

Avoid the four deadly sins of apple growing

1. <u>Too low a soil pH.</u> The optimum pH of the soil is 6.5 to 6.7 for apples, as a rule. In drizzly winter areas such as mine, soils leach out calcium and we frequently find the pH too low, often between 5.0 and 5.8. Critical minerals are then either unavailable or in very short supply to the tree roots. This can contribute to bitterpit and other internal breakdowns of the apple.

2. <u>Excessive use of nitrogen and other fertilizers!</u> This results in apples frequently too large, with lack of sound cell structure from a shortage of calcium in the apple. Too much of the calcium is going into the rampant growth of limbs and leaves, brought on by over-use of nitrogen. Storage time would also be reduced.

3. <u>Picking the fruit much too early</u>, so that soluble solids (sugars) are not high enough and the fruit is starchy and bland.

4. <u>Storing an apple way beyond its natural storage time.</u> When it comes out of storage, it goes downhill rapidly and becomes mushy. We are trying too much to make some apples fit a preconceived marketing program instead of making marketing fit the genetics of a specific cultivar. Fuji is one of the few cultivars that will take considerable room temperature abuse (up to 3 weeks or more) and emerge crisp, juicy, and firm. Granny Smith and Newtown are close to that.

Quality ratings

This book renews a time-honored system for describing quality, which is primarily an evaluation of flavor. Keeping quality and annual bearing are also important traits. This system was brought to its peak by Beach, Taylor and Booth in their two-volume classic *The Apples of New York*. In 1857 Hooper used ratings of 1, 2 or 3, which help, but are not as accurate as Beach's system.

Apples with two of Beach's ratings are left out: "poor" and "fair" in quality, because no apples of quality that low are given complete descriptions in this book. Because *The Apples of New York* was completed in 1903 and published in 1905, none of the newer cultivars were known and the ratings for those are mine. I hasten to add that these newer ratings are often based on over 11,000 taste tests involving the general public and are only in part my own opinions. None of the ratings were made in haste. The quality ratings are as follows:

Poor or **Fair**: No cultivars of this low a quality are described, although some perhaps fall in the quality of "fair" in USDA zones 3 or 4; however, those would be mentioned because of great hardiness.

Good: This is an average quality, and only a few are described in this book. Cultivars rated "good" are included only because other, better-tasting cultivars do not grow where one rated "good" will, usually because of climates such as in USDA zones 3 and 4 in the north, or zone 9 in the south.

Good to Very good: Above average quality. In most instances, this is the lowest rated cultivar in this book. The old Baldwin was so rated by Beach. Baldwin was strictly for cooler climates and was often rated as poor farther south, and is not included in this book.

Very good: A great many cultivars described are of this quality. Apples so rated have an above average flavor and usually other desirable characteristics.

Very good to Best: Only a few apples available worldwide deserve this rating, perhaps 25 to 40 or so. Examples include properly ripened Golden Delicious, Northern Spy, Spitzenberg, and the newer Jonagold, Braeburn, and Gala. I think Fuji is such a Jekyll-and-Hyde type I would not rate it this high.

Best: In *The Apples of New York*, Beach rates only two cultivars in this rarefied status. One is Newtown, a rating with which I agree because Newtown has many fine characteristics. The other so rated was Summer Pearmain, a very obscure August-September cultivar requiring about 2 months of selective picking, which is unusual but apparently is of very choice flavor. (It may be hard to find because more than one cultivar was so named. Which is which? How do we know?) Great care should be taken in placing a cultivar in the "best" category. Few apples could be so classed without eliciting considerable controversy, because there is a fine line between "very good to best" and "best." A "best" rating should probably only be given after an apple has been tested over at least 25 to 50 years in a wide variety of areas across North America, and indeed the world. I know of no cultivars that I would add to "best" yet; perhaps a few years from now!

Researchers and magazine contributors often write that an especially good apple is "excellent" in quality. I would rather follow this more exact system.

Why the Apple Cultivars in this Book?

The question opening this section may imply there are hundreds of above-average apple cultivars excluded, but this is simply not so. Fifty apples are described at length in this book, with color illustrations, and another 40 or so are mentioned for the northern zones of 3 and 4 and the southern latitudes of zone 9. A few good-eating or long-keeping cultivars were excluded either because of poor tree characteristics or too much similarity to others that are described, and a very few have no complete information available. I have grown, tested, and photographed all 50 that are fully described, except for Wickson crab. This book, then, is primarily about *my* work, in *my* climate.

I think Jonalicious is one of the best examples of a fine-tasting apple, but with some undesirable tree characteristics. I only know of one direct marketer, in California, growing it commercially in the West. Along with Criterion, it is his best seller because of very fine flavor, but both have difficult tree characteristics. In any event, it has not found favor here although it was discovered in Texas in 1933 and has been available from Stark Brothers since 1958. The main reasons it is excluded are its similarity to others described here that I think are much easier to grow, and due to its having a cullage rate of 30 percent or so. Jonalicious apples are available for tasting by mail from "Apple Source."

The influence of fresh apples shipped from New Zealand into British Columbia, Washington, Oregon and California has accelerated the testing of certain cultivars described at length in this book. Since our growers have done a great deal of travelling to Japan, New Zealand, Australia, South America and Europe, testing will be accelerated on some of the newer cultivars just now becoming available, such as Pink Lady, Sundowner, Honeycrisp, etc. Virus-tested trees are now available in limited quantities. Testing will take additional time.

An acquaintance, who probably should know better, said that no matter what cultivars were in this book, they would be outdated by new ones in two years. This person probably flunked history in school. For example, the three hottest new commercial apples being planted in the U.S. from 1990 to 1994 were Gala, Fuji, and Braeburn. They were crossed or found in 1934, 1939, and about 1949, respectively; that is an average of 54 years, and many of our citizens have probably not tasted them yet. The next group of new cultivars won't take that long to be established because of wider testing and grower travels, but it will take much longer than two years.

Almost 100 years ago Beach, in *The Apples of New York*, rated Evening Party and Newark Pippin, among others, as having "very good to best" flavor. Both had tree and fruit defects and apparently are lost to culture, at least here in the west. Their good flavor did not save them, as apple cultivars must have other good characteristics to survive over time. So, best they remain dead and buried along with many others of their ilk.

An enormous number of cultivars have been derived from the species of apples in the genus *Malus*. By 1900 over 5,000 had been named in the U.S. in the previous 250 years, many of questionable value. When inferior apple cultivars disappear, only a strain dies, not a genus nor even a species. We need not mourn their passing, as though we had just cut down the last redwood tree. Let us invest our energies selecting the best apples, whether new or old, and then some will survive for centuries while new and better ones may come along.

After I had wasted too many years testing inferior cultivars, a simple formula reared its head to stop my insatiable desire to grow everything ever named: if it is an old type and neither Beach nor Hooper gave it a high rating, I do not plant it now. For new apples, go to fruit shows in your area or to a good direct marketer and taste the apples. "Apple Source" now allows you to taste many, old and new, by mail. I have had to use trial and error with the new cultivars because I try to stay 5 or so years ahead of the crowd, so accurate advice can be given to direct market growers with whom I work. I think this book will help you select both old and new apple cultivars. Yes, there are only about 100 mentioned and only 50 or so described in detail. Even that many may be confusing, and not all will grow and fruit well in every climate. But then, why would anyone want 400 to 500 cultivars in one's back yard or direct-market farm?

Hooper's *Western Fruit Book*, written in 1857, has a 1-2-3 rating and is instructive, even though the "west" was a warm, humid Ohio. Hooper rated highly such fruit as Ralls Janet (one parent of Fuji), Newtown, Spitzenberg, White Winter Pearmain, Northern Spy, and Fall (Fall Pippin). This book is extremely rare, but the two volumes of Beach's *The Apples of New York* can still be found, and the various fruit societies usually have copies for study. Public libraries seldom keep such old books and are usually a poor source. However, these volumes are wonderful reference and historical books. Beach was asked to include all these cultivars in his report of 1905. In my opinion, probably less than 75 are now worth growing—or even available. Perhaps no one then could foresee the great advances in refrigeration and transportation which were soon to occur. As you read this book, you can understand my different approach, which is more in touch with a population now 97 percent urban.

Those who wish to experiment with old cultivars not listed in this book could write to such nurseries as Southmeadow Fruit Gardens, Greenmantle, Sonoma Antique, Bear Creek, Rocky Meadow and Living Tree Heirloom Apples (see "Nursery names and addresses"). Unfortunately, most descriptions are meager and do not usually list the bad traits of each cultivar. One needs to fruit them for years to determine their true traits in a specific climate. This is not a short-range hobby or profession.

In more recent years I have moved from being a collector to a selector of apple cultivars, and this is the main ingredient of my formula. Some examples: One acquaintance asked, "Do you have Thompkins County King on your list?" No, not in the 1990s. King does have fine flavor, but it is a triploid, is often a sparse producer in poor pollinizing weather, the fruit commonly develops water core, is not very hardy, has a very greasy skin, is frequently too large, and develops bitterpit. I grew King for many years but today it seems better to grow Melrose and Jonagold. I think Jonagold tastes better, and Melrose is a far better keeper and tastes as good. Unlike King, I have found both to be dependable bearers in my climate.

Another example of selection at work is making a choice between similar cultivars such as Karmine de Sonnaville and Holstein. Both are from northern Europe, both have Cox's Orange as one parent and both are triploids. Karmine™ has even better flavor and is easier to grow. The tree is smaller than Holstein and its apples seldom crack, as with Holstein, after the first year of fruiting. Holstein is now just a pleasant memory. Such a selection could not have been made unless I had grown both of them. Having a somewhat limited space has helped me to select and remove.

A much more difficult decision was to scratch Golden Russet from this book's list. Ashmead, in my opinion, has a more sprightly flavor. Hudson Golden Gem is a large, colorful, full russet with a highly acceptable pear-like flavor and has resistance to the apple scab fungus. The old Roxbury Russet is the oldest U.S. apple of note, dating to before 1649. It is somewhat self-fertile, an excellent mid- to late-blooming pollinizer, and is useful for juice, sauce, fresh dessert and drying. The tree is very moderate in size and is resistant to apple scab. So, these three russets were selected from about 12; however, there is no law that says one cannot grow others. Unfortunately, some are difficult to secure except perhaps from Southmeadow Fruit Gardens. I just don't want that many russets or I would be succumbing again to my old addiction. "Apple Collectors Anonymous" is a new club, and joining it gave me peace of mind and a lot more free time.

Another good example where selection is at work is in Japan. About 12 years ago some of our growers and researchers came back from there with glowing reports of at least 20 new apple cultivars that were to take the world into a new age. These same people were there recently, and now these fantastic new cultivar numbers have shrunk to about six or so, and they may not be better than what we are growing. Some are just now being tested in the U.S. and are years away from fitting into our various climates, if they ever do.

I think you can be assured that 40 new apple cultivars will not suddenly upset our whole system and replace many thousands of acres invested in current cultivars, commercial and private. Cost alone would prohibit it, and when consumers get attached to certain cultivars and their uses, they will continue buying them and/or growing them in their back yards. Many cultivars cannot be purchased in stores and, if you want the best tasting apples, you may need to grow at least some of them yourself.

To be given a complete description in this book, I require that they usually fit the following guidelines:

- Grafted trees should be available for purchase in the U.S. from at least one nursery by 1996.

- The cultivar should be legally available only after having been disease tested by one of our two USDA testing stations: at Prosser,

Washington, or Beltsville, Maryland. Many foreign apple cultivars have viruses or other diseases that could threaten our current production, especially the vast acreage of Red Delicious in eastern Washington State, of about 110,000 acres. Some private testers do not understand the importance of these agricultural laws, and may be growing infected trees. Even if for their own use, this is not advisable and in many cases is illegal unless under special license (which some testers have).

- If the cultivar is in the patent pending process, or has a patent number, these trees should not be offered for sale except by those licensed to do so. There are some new Japanese cultivars that may become licensed in the U.S. and are not yet legal to grow here, so no complete descriptions will be given—although rumors of their quality may be mentioned.

All apples are not given complete descriptions or pictures but are mentioned as desirable to grow or test. (See the chapter titled "Notable New Apples.")

The Japanese are developing six or seven fine new cultivars, but they cannot be given descriptions in this book because so little is known about them. The general availability of these trees in North America is perhaps four to 10 years away.

Climate Effect

Certain apple cultivars seem to produce and fruit almost to perfection in one area, and are regarded as of less quality in other areas. Usually, pre-harvest heat is the main reason. If daytime temperatures are consistently in the 85° to 105°F range for about three weeks before harvest, only a few apple cultivars will emerge from this furnace without sunburn, bitterpit, shorter storage time, etc.

Cool pre-harvest nights, many in the 40°F range, are what probably save eastern Washington later apples from the rather high daytime heat. Red Delicious seems especially adapted to this area, which has the highest concentration of this cultivar in the world.

Some skillful growers in hot areas are growing average quality Jonagolds if the area is above 1,500 feet altitude, has automatic overhead tree misting systems that turn on when temperatures reach about 85°F, and a number of calcium chloride sprays are applied. In any area where growers have to pick Jonagold before October 1 to 10, the cultivar will have lesser quality and keeping characteristics. Our Willamette Valley is a much better area in which to grow Jonagold at lesser cost. Washington and Oregon, west of the Cascade Mountain range, have among the best climates worldwide in which to grow Jonagold. Growers in hot areas should select an apple cultivar more suited to the climate.

Ripening mid-season—August 10 to September 20:

Arlet (Swiss Gourmet™), Ginger Gold™, Gala, Gravenstein, Elstar.

Ginger Gold™ and Gala do not follow the rules very closely, as they tend to give fine quality in most climates and stand rather high pre-harvest temperatures, especially if there are some cool pre-harvest nights. Gravenstein and Elstar demand a cooler pre-harvest area such as here in the Willamette Valley. All five do well in our climate, with clay-loam soils. I want nothing earlier than this group because I think earlier ones are of lesser quality and of only hours to days in keeping. I no longer grow any first-early apples.

Ripening mid- to late-season—in western Oregon, September 20 to October 15:

This group is somewhat more difficult to assess regarding pre-harvest temperatures. Remember, there are microclimates within microclimates and each year is not the same (especially 1991 and 1992.) Mt. Pinatubo in the Philippines had a cooling weather effect on the Northern Hemisphere during that period. El Nino also had a strong effect, especially here in Oregon. In 1992 we had the only warm and dry March and April I can remember in 31 years. Chill time in 1992 had been met by Christmas for all fruit, and most of the apples, cherries, and pears bloomed at the same time. Everything ripened about two weeks earlier, because bloom was two to three weeks early.

Statistics from the U.S. Weather Bureau in Portland, Oregon, show that the average number of days per summer that reach 90°F or more between May 15 and October 15 is 10.7, so this has to be regarded as a fairly cool area, with most of the hot days coming in July and August. Cool nights start in late August or, at least, by September. This is an ideal climate then for Jonagold and Braeburn. Both get bitterpit badly in too hot a climate and, even here, commercial growers usually spray calcium chloride many times to help prevent it. Home growers here generally do not spray calcium chloride because of our favorable climate and irrigation.

<u>Will take a warmer area—ripening period September 20 to October 15:</u>

Ashmead, Blushing Golden™, Red Delicious, Freyberg, Golden Delicious, Grimes Golden, Haralred, Jonathan, Keepsake, Spigold, Wickson Crab, Orleans, Kandil Sinap.

<u>Prefer a cooler and/or more northern area—September 20 to October 15:</u>

Bramley, Cox Orange, Cortland, Empire, Erwin Baur, Hudson Golden Gem, Jonagold, Karmijn de Sonnaville, Kidd's Orange Red, Liberty, McIntosh, New York #429, Rhode Island Greening, Sweet Sixteen.

Long ripening season—October 15 to November 15, at least in my climate:

In warmer or cooler areas, these apples need a long fall with no heavy freezes before mid-November. They seem more climate susceptible than any of these groups listed. Most do best in USDA zones 6 through 8, with trial in lower zone 5 and upper zone 9.

<u>Long growing season (but warmer):</u>

Fuji, Granny Smith, Stayman, Sturmer, Tydeman Late Orange, Winesap, York.

<u>Long growing season but prefer a cooler long season:</u>

Braeburn, Calville Blanc D'Hiver, Coromandel Red, Idared, Melrose, Mutsu, Newtown, Northern Spy, Orleans, Roxbury Russet, Spitzenberg.

Where are these growing areas? With such diverse climates as we have in North America, the answers are difficult to come by. You will find only USDA zones and general guidelines here; however, that is more than is usually presented, especially for the newer cultivars.

Where will an apple tree grow and give edible apples? Almost anywhere in North America where it is not too cold in winter, there is sufficient chill time for a certain cultivar, and a long enough growing season for that cultivar.

In 1975, I was asked to escort a young Guatemalan around my area and visit a number of nurseries. He said he lived at only about 14 degrees above the equator in what he called the highlands of Guatemala, and he was very interested in buying many thousands of dollars worth of fruit trees, especially apple trees. He had a nursery and an apple orchard of some commercial value. Around him, primarily in southern Guatemala, are some very high mountains, a few 11,000 to 13,000 feet high. His orchard was between 8,000 and 9,000 feet of altitude, featuring mostly the cultivar Anna from Israel but also Red Delicious and Golden Delicious.

As one goes up in height, the climate becomes somewhat similar to that of a more northern latitude. Ultra-violet rays are more intense at higher altitudes, which have some effect on bloom, ripening, etc. Theoretically, for every 1,000 feet in additional altitude above sea level, the temperature is about 3°F cooler. The above example would be a southern extreme and only then at higher altitudes. Unfortunately, I did not follow up on this anomaly and cannot provide any real conclusions. Mexico has considerable apple acreage and much spring frost at altitudes where they are grown.

Then there are the northern limits such as Finland at 60 degrees north latitude, where apples are grown commercially near large bodies of water with a short season and also relatively low chill cultivars. Acquaintances in Anchorage, Alaska, near 61 degrees north latitude and near water, are testing various cold-hardy apple cultivars on cold-hardy rootstocks. While in Anchorage in the fall of 1976, I was asked to identify an apple that was about 25 years old. It was a Westfield Seek-No-Farther, an old apple which has fine flavor and, according to Beach, is very hardy and of "very good" flavor. Rootstock was unknown but thought to be one of the crabs. One of the problems in much of Alaska is permafrost a few feet down.

The above extreme areas are not prime growing areas, but accentuate some of the climate diversity of this remarkable fruit called apple.

Since we have separated the cultivars into three sections, with further separation into cooler and warmer areas, perhaps some suggestions could be offered that would be helpful, at least for the cooler areas.

I was discussing this with a knowledgeable friend, only half seriously, and told him I had devised a simple plan for the eastern U.S. Start at Davenport, Iowa, on the Mississippi River, and draw a line east about 25 miles south of Chicago directly to Boston, Massachusetts, on the Atlantic Ocean. Anything north of that line is a cooler climate; everything south of it is a warmer climate. This is a fairly acceptable idea, but then what about the difference when in the mountains of the Virginias or North Georgia, or on the east side of a good-sized lake? or..., or..., or, ad infinitum? At 2,000 to 3,000 feet in the Virginias, there is also prime apple area.

The best answer is to try to find someone in your vicinity that has successfully grown the cultivars you are interested in. The newer cultivars, of course, have not been tried in some areas, so go to your local Extension Service sources and also use the information in this book with the individual descriptions. Then test your area yourself to be sure.

There are some obviously cooler areas, such as in Michigan, one of the top three apple production states. With a long large lake on its west side and the general westerly winds riding east across the state, there are fewer spring frosts or early fall freezes. Large bodies of water do not change temperature nearly as rapidly as land does and even a small lake 2 or 3 miles across can mean that growing apples on its east side would be affected. The Door Peninsula in Wisconsin juts up between Green Bay and Lake Michigan proper and is quite cool most years. That area grows some fine apples which have been carefully selected for a cooler climate.

There are gamblers in every profession, and some go into very expensive apple plantings with no testing by anyone in that area. In southern California, the hot valleys are a prime example. Substantial plantings of Braeburn went in with insufficient or non-existent testing and, because of severe bitterpit and internal browning, these trees are being removed. Now some are ready to gamble with other untried cultivars. Wouldn't it be better to test a cultivar for a few years rather than be one of the first to flirt with bankruptcy? And what about their management? Did they take an acre with automatic cooling misters and spray 10 to 15 times with calcium chloride? Did they consider the other factors contributing to bitterpit? Can they blame it all on heat? I don't know for sure!

Obviously, I have gone to considerable length to stress that even a small commercial orchard should be researched before planting. Commercial apple plantings can be expensive.

Consider cost: If one has to buy the land, drill a well, and hire labor, it might be $10,000 or more per acre. These costs certainly warrant few mistakes.

Looking at other orchards in your climate is perhaps the first step in research. If the answers are not there, then you must test the various cultivars yourself when very new cultivars are involved, although the above guidelines have been helpful to those with whom I work. Our main problems in this valley, though, are usually not our summer climate but rather our willingness to control apple scab fungus, powdery mildew fungus, codling moth and apple maggot fly. If control is handled properly, we grow some of the best quality apples in the world and sell them at premium prices.

Conclusion:

Let us be careful about calling a certain apple a "bummer" in one climate, even though we may know it has proven quality in another climate—perhaps not too far away. Growers in hot areas have called such cultivars as Elstar, Cox Orange, Gravenstein, Jonagold, Melrose, and Brock, unfair derogatory names even though they are superb in cooler areas. That's why there are many different apple cultivars—so we can find at least some that thrive in a specific area, especially the area where you, the reader, lives.

USDA Climate Zones

After looking at apple orchards from coast to coast, I came to the conclusion that the thousands of microclimates in North America could not be clearly defined, so I will use the USDA zones 3 to 9 as a general guideline, however imperfect that system may seem to the reader.

More apple cultivars will do better between zones 5 and 8 than in zones 3 and 4 or zone 9, but perhaps as many as half of the 30 commercial cultivars described in detail will do well one zone farther north than suggested, if planted in favorable microclimates, such as on the east side of a large lake.

Please remember, growing apples in any climate is not an exact science.

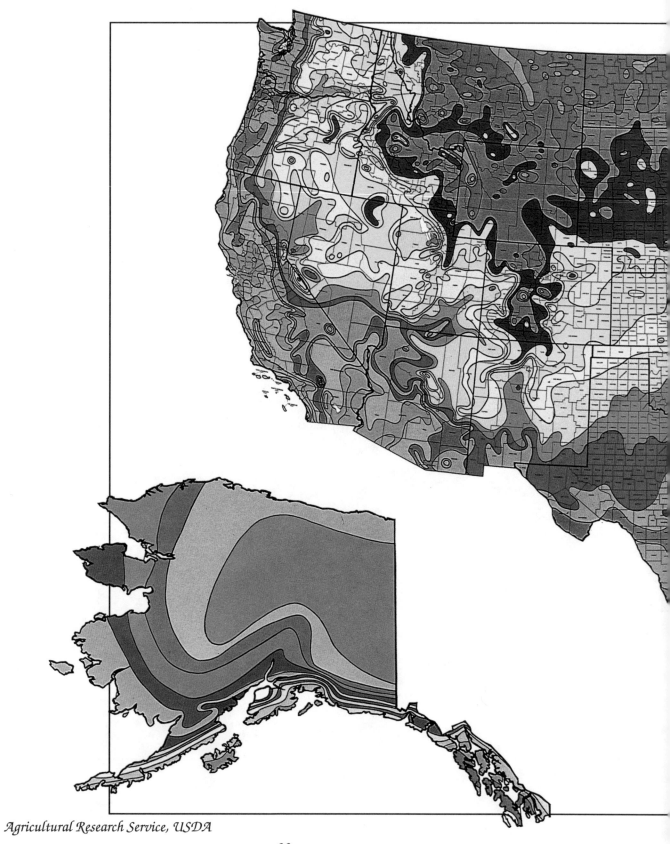

Agricultural Research Service, USDA

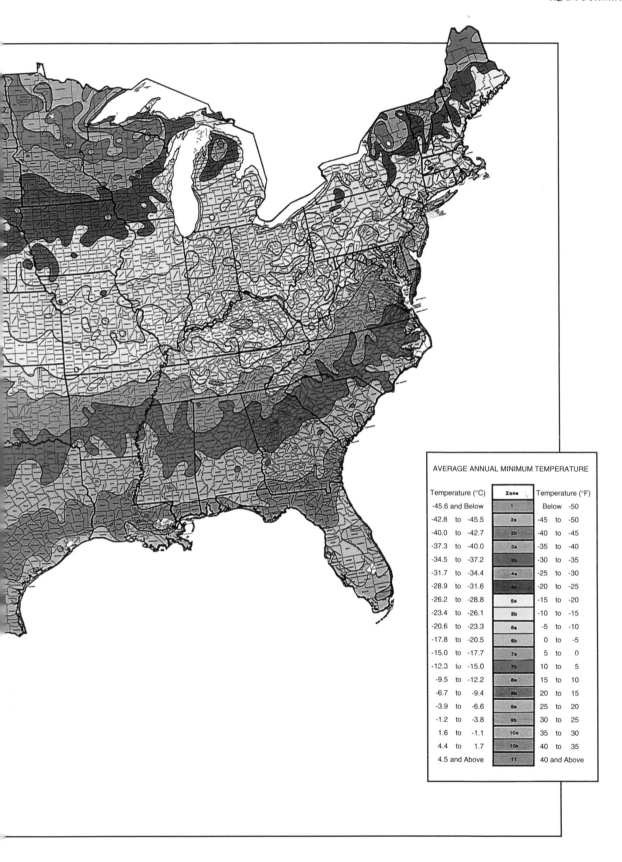

AVERAGE ANNUAL MINIMUM TEMPERATURE

Temperature (°C)	Zone	Temperature (°F)
-45.6 and Below	1	Below -50
-42.8 to -45.5	2a	-45 to -50
-40.0 to -42.7	2b	-40 to -45
-37.3 to -40.0	3a	-35 to -40
-34.5 to -37.2	3b	-30 to -35
-31.7 to -34.4	4a	-25 to -30
-28.9 to -31.6	4b	-20 to -25
-26.2 to -28.8	5a	-15 to -20
-23.4 to -26.1	5b	-10 to -15
-20.6 to -23.3	6a	-5 to -10
-17.8 to -20.5	6b	0 to -5
-15.0 to -17.7	7a	5 to 0
-12.3 lo -15.0	7b	10 to 5
-9.5 to -12.2	8a	15 to 10
-6.7 to -9.4	8b	20 to 15
-3.9 to -6.6	9a	25 to 20
-1.2 to -3.8	9b	30 to 25
1.6 to -1.1	10a	35 to 30
4.4 to 1.7	10b	40 to 35
4.5 and Above	11	40 and Above

My Test Orchard Microclimate

Before proceeding, I must discuss the unusual climatic conditions my test orchard enjoys, lest I draw criticism from orchardists in areas where fewer cultivars can be grown due to harsher growing conditions. I grew up in southern Minnesota so I well know the limitations of a shorter season and a sometimes severe cold winter with a high wind chill factor.

I have grown and tested 136 cultivars in Oregon in the last 31 years. These cultivars originated in 12 countries worldwide. My orchard is primarily a test orchard, but it also supplies my family, relatives and friends with fresh dessert, cider, sauce and apples of the very best flavor for baking. I test primarily for taste, annual bearing, disease characteristics and storage life.

This small orchard also has included apples of historical significance and high quality such as Spitzenberg, Northern Spy, Calville Blanc d'Hiver, Newtown and Ashmead. It also contains a number of newer foreign high-quality apples such as Fuji, Elstar, Gala, Braeburn, Honeycrisp and others.

Fortunately, there is only one factor I cannot test for—cold hardiness. Incidentally, I believe most of the two northern tiers of states and the southern reaches of the Canadian provinces could grow a wider assortment of apples if they planted the hardiest cultivars on hardy rootstock, protected the orchard with a windbreak of coniferous trees on the north and west, and supplied supplemental irrigation for years of drought.

What follows is not a weather report, but a description of my microclimate to help the reader understand why I can obtain good quality apples from cultivars developed or discovered over a wide range of climates and regions. My small orchard is 10 miles south of downtown Portland, Oregon, and is planted on a hillside at about 300 feet altitude, one-half mile from the Willamette River. It is a cooler, mild type of USDA zone 7 to 8, and is quite unlike zone 8 in east Texas, which tends to be humid and warm.

1. I receive approximately 2,500 or more hours of chill time between 32°
   to 45°F most years which is far more than most cultivars need, as I
   know of no apple trees requiring more than 1,400 or so.

   Dormancy is an annual period when most physiological factors come
   to a halt, but apple trees go through two types of non-growing rest:

   (a) chill time, a period with temperatures between 32° and 45°F, and
   there is some disagreement on these figures but we do not fully understand the physiology of trees.

   (b) a true dormant period when the temperature may be considerably
   below 32°F, sometimes below 0°F in the northern areas of the U.S.
   and southern Canada, and above 45°F in the south. Regardless of
   how low the temperature drops, if the tree does not experience a
   sufficient number of chill hours it blooms erratically if at all. (See
   the "Chill Time" chapter.)

   The middle latitudes of the United States, with their milder winter
   temperatures and considerable temperature fluctuation, uniformly require high chill cultivars so that the trees do not break dormancy and
   bloom in a warm period, which sometimes occurs in late January or
   February. These cultivars are primarily grown in USDA zones 6 to 8.

2. Sparse spring sunlight, cool temperatures and a drizzle keep my trees from
   blooming too early in the spring, even though chill time has been met.

3. I have never lost a crop in my microclimate to late spring frosts or early
   fall hard freezes. My small orchard is on a hillside surrounded by dark
   colored fir trees, so the orchard tends to be 10° to 15°F warmer than the
   open flat areas of the valley during fall and spring cold snaps.

   I have picked very late varieties such as Granny Smith, Newtown,
   Fuji, Sturmer and Braeburn from November 5th to the 15th, when the
   sugar content was optimum for eating but probably slightly late for long
   storage. In short, the orchard has the advantages of a long, mild fall
   and a long growing period.

4. Westerly winds off the Pacific Ocean, 80 miles away, usually keep the
   summer smog to an acceptable level, so the trees enjoy good light inten-
   sity. The latitude of the orchard is about 45 degrees north, the same as

Halifax, Nova Scotia, and St. Paul, Minnesota. This latitude provides long daylight hours for maximum photosynthesis. Over half the orchard receives sunlight even at 8 p.m. in May, June and early July.

5. Daytime temperatures are seldom above 90°F, although they occasionally go higher for a few days in July or August. Cool marine air seeps into the valley at night, breaking summer hot spells, so daytime temperatures are typically in the 70° to 85°F range. As a consequence, sunburned spots on susceptible varieties such as Granny Smith, Bramley, Baldwin, Newtown and Jonagold are seldom seen.

   The pre-harvest night temperatures in late August, September and October are typically in the 35° to 55°F range. These cool night temperatures are important for the formation of highly colored apples. Because of cool pre-harvest nights, I can usually grow an acceptable McIntosh.

6. The summers are rather dry, so irrigation is necessary for consistent crop volume. Rainfall at my orchard averages between 32 and 40 inches per year, most of it from seven months of drizzle from October through May. The dry summers mean little cloud cover, so scab sprays are usually unnecessary after mid-June, assuming the trees were diligently sprayed from blossom time until the rains stop in May or June.

7. One other important fact is that there is almost never fireblight on apple trees in the valley floor. Its absence is of great importance for both apple and pear production. The bacterial disease fireblight, when present, will limit production and could kill trees. Also, we do not have the obnoxious pest plum curculio.

These seven factors are good news. The bad news is that there is apple scab, powdery mildew, anthracnose, codling moth and the recently introduced apple maggot fly. Fortunately, these disease and insect problems usually can be safely controlled at minimal cost. The Golden Delicious cultivar sometimes russets due to our drizzly spring weather or mildew disease. Red Delicious strains are so scab-prone that they require special care, but that is of small importance as these two cultivars are usually not recommended here because of high production from dryer climates in other areas, especially eastern Washington.

My description would be incomplete if I did not mention that the Willamette Valley of Oregon is a cool, green, lovely place in which to live.

Chill Time

This subject was mentioned earlier but needs reference to specific cultivars because both home orchardists and small specialist orchardists across the continent are now trying to grow apples, often in areas previously thought to be marginal.

All apple cultivars require a certain chill time between 32º and 45ºF, but some are classed as "low chill" and do not require very many hours each year in the above temperature range. If the trees and buds do not obtain enough chill time for that cultivar, the tree blooms erratically or not at all. Interestingly, the climates of not only the southerly regions but also the more northerly regions often do not provide sufficient chill time for many apple cultivars; in the south, because the temperature in the winter is often above 45ºF and, in the north (USDA zones 3 and 4), the temperature during dormancy is often consistently below 32ºF. These are not prime apple growing areas except for certain cultivars on appropriate rootstocks.

Scientists do not study the low-chill cultivars much because they are grown outside of prime commercial growing areas and are often not regarded as commercial. Testing done by amateurs, then, is important.

Listed below are some better quality low-chill apple cultivars requiring chill time of only 300 to 700 hours or so, and most are being tested by home orchardists, many in the Los Angeles basin. If one plants in the foothills in the south at 1,500 feet or more, obviously chances for fruit are better. Each 1,000 feet of altitude lowers air temperature an average of 3ºF.

There are a few of the cultivars mentioned below that are also cold-hardy to USDA zone 5 and in protected areas of zone 4. They are Gala, Gordon, Winter Banana, White Winter Pearmain, Yellow Bellflower, and perhaps others not sufficiently tested.

Most of the good tasting crab apples are low chill, and they usually grow in the south or the north. My favorites are Wickson, Centennial, Dolgo, Young America and Whitney.

Consider joining the North American Fruit Explorers, as the information they provide at meetings and in the quarterly publication *Pomona* is most helpful. The Home Orchard Society is another good source of information, and they also have a quarterly publication. (See Bibliography.)

Low Chill Cultivars
(Approximately 300 to 800 hours needed)

1. Newtown	11. Yates
2. Pink Lady	12. Mollie's Delicious
3. Sundowner	13. Yellow Bellflower
4. Stayman	14. Gordon
5. Golden Delicious	15. Winter Banana
6. Gala	16. Beverly Hills
7. Fuji	17. Anna
8. White Winter Pearmain	18. Ein Shemer
9. Hudson Golden Gem	19. Dorsett Golden
10. Kandil Sinap	20. Stark's Adina

Apple Chromosomes

Apples are in the family *Rosaceae* and the sub-family *Pomoideae*. The genus is *Malus*. Of the at least 14 or 15 apple species in the world, one, *Malus pumila* of Europe and southwest Asia (Iran, Turkey, Russia), probably is an ancestor of our common apple.

Apple trees have 17 chromosomes in a set. Most commonly grown apples are diploids, meaning that they have two sets of chromosomes, equalling 34. Except for Winesap, all commonly grown diploids have viable pollen for the cross pollinizing of other apple cultivars. Winesap is a diploid that has very defective pollen and should not be used to cross pollinize. In orchard placement, treat Winesap as a triploid (three sets of chromosomes).

Triploids, $3 \times 17 = 51$ chromosomes, are our main interest in this section, as they have defective pollen and should not be relied on to cross-pollinize themselves or other cultivars.

The triploid, with its odd number of chromosomes (51), does not divide evenly during cell division, which results in defective pollen. On the other hand, these triploids almost all have some self-fertility, with Bramley, King, and Stayman perhaps having the most. It is not enough self-fertility to give very much fruit, and it should not be relied on. To my knowledge, the process of fertilization is not completely understood. The important factor is that the triploid be identified.

The following triploid cultivars are described at some length in this book. Their chromosome count of 51 has been confirmed by a number of research stations: Bramley, Karmijn de Sonnaville, Stayman, Gravenstein, Mutsu, Spigold, Jonagold, Rhode Island Greening.

There are other cultivars not described here that are still grown in the U.S., or are grown to a limited extent in Europe and should also be identified as triploids: Baldwin, Holstein (new), Ribston, Belle de Boskoop, King, Suntan (new), Blenheim Orange, Reinette du Canada.

Almost without exception, the triploid cultivars produce large apples and, without superior orchard management, they are quite subject to bitterpit, especially on or from young trees.

Jonagold is an appropriate triploid to use as an example, as in the 1980s it was the most planted new apple in northern Europe and west of the Cascade Mountains of Oregon and Washington. It is a cultivar needing the cooler late summer climates.

Only five years ago, one commercial grower planted just one diploid with his Jonagolds and wondered why he got such skimpy crops in the third and fourth leaf, despite heavy bloom. Neither cultivar had received adequate pollinizing. With Jonagold and other triploids, you should also plant two diploids, both blooming with good overlap to pollinize each other and the triploid. Many growers in my area like to use Akane and Melrose as they are also very saleable cultivars. Although blooming with good overlap, the above three cultivars ripen in a sequence from about August 25 to October 25, giving a long sales season for the small commercial or home orchard.

Apple Blossom Pollinization and Bees

The matter of apple pollinizing is, for some reason, largely ignored by home orchardists and often incorrectly reported by writers. All commonly grown diploids are adequate pollinizers. The exception is Winesap, which has defective pollen. Another potential problem is using parents to pollinize first generation offspring. This is not a hard rule but should be avoided if possible; however, it is a hard rule with Jonagold and one parent, Golden Delicious. They are not cross-fruitful.

So now there are a number of factors to understand to ensure maximum cross-pollinization.

- Parents of a cultivar, if known. (See descriptions.)
- Bloom periods, for overlap. (See chart. Please make your own bloom chart.)
- Which cultivars are triploids or diploids. (See the preceding "Apple Chromosomes" and the individual apple descriptions.)
- USDA zones potentially suitable for a cultivar's success. (See descriptions.)

It is futile to buy trees and prepare soil, perhaps put up posts and wires for a trellis, and then create an environment where cross-pollinization is poor and crops are inadequate.

In my domestic test orchard there are from 35 to 50 different cultivars undergoing testing in a given year, growing in trellised rows with trees from 2 to 10 feet apart. The bees then find pollen available in quantity for all bloom periods, and they tend to fly down the rows rather then the greater distance between rows.

Very late and very early bloomers, however, did not always give full crops until I planted them next to a different cultivar of the same bloom period, the closer the better. Ten or 15 feet apart for pollinizers is ideal. If 100 to 200 feet apart, fertilization goes down.

The *pollinator* is an organism that transfers pollen from one flower to others. The cultivated honeybee, various wild bees, the big bumblebee, and certain bee-like flies are the common pollinators of apple blossoms. Humans also do hand pollinating for cross-breeding, with pollen from a known cultivar so parents are properly identified.

The plant supplying the pollen is called the *pollinizer*. All commonly grown diploid apples (2 x 17 = 34 chromosomes) produce viable pollen except Winesap.

Some researchers have shown that when the same flower was visited by pollinators at least six times, there were more seeds formed and better fruit set, with higher calcium levels in the apples.

All apples should be treated as self-sterile, even though some exhibit limited self-fertility. Cross-pollinization always produces larger crops.

A nursery in the eastern United States contends that the over 100 apple cultivars they sell are self-fertile and can be planted alone; however, they also state that two apple cultivars planted together will give more fruit. This over-simplification can be quite misleading. Suppose one is a triploid? Suppose one bought an Idared, a very early bloomer, and a Northern Spy, which is a very late bloomer? In the northern U.S., where spring may come suddenly and stay warm, apples have considerable bloom overlap. In my climate Idared drops its blossoms before Northern Spy blooms are open and there would be no cross-pollinization between them.

The bloom chart in Part III is based on a six year record I made for western Oregon and western Washington in the USDA zones of 6, 7, and 8, but primarily zone 7 (mine). It will differ, where apples can be successfully grown, in the far north and far south of North America.

Our bees

The cultivated honeybee, handled by beekeepers, and our wild bees are responsible for pollinating about 95 to 100 crops in North America. Their value is many billions of dollars per year, but most of us have taken them for granted.

They just appear every spring and do their job of seeking nectar and pollen from flowers. While doing this, they transfer pollen from one flower to another, which is the first stage leading to fertilization and full crops of apples.

In the 1980s bee colonies began to be invaded by two very different parasitic mites from other continents. These organisms pose a serious threat to bee populations.

One of the invading organisms is an endoparasite of microscopic size; it is called the tracheal mite and as many as 50 or 60 can reside in a bee's trachea at one time. This gives us some idea of how small they are. The tracheal mite has invaded cultivated bee colonies in over 25 states. The tracheal mite is apparently an invader from Europe, where it is indigenous. It was first seen in England around 1904. Some of these English strains of bees now have resistance or at least tolerance to this mite, and queen bees are being imported from England in hopes that this tolerance is passed on genetically to our bees. There is already some evidence that bees are tolerant when bred from queen bees fertilized by tolerant males. Menthol, a readily available product, is being tried as a control after honey has been removed. Its long-term effectiveness, I believe, is unknown as of this writing.

The other mite is an ectoparasite called the Varroa mite. It apparently has invaded the U.S. and Canada, probably from eastern Europe, but it came to eastern Europe in the early 1900s from the Ukraine and farther east in Russian Asia. This mite feeds on bee blood externally, and an adult mite can be seen on the bee if one looks closely.

The Varroa mite is very deadly and control is being tested with pyrethoid pesticides. This class of pyrethoids is purported to have low toxicity to humans. Its effectiveness and legality in various states are unknown to me. If untreated, the Varroa mite will kill every hive it infects within a few years. It was detected in southern Oregon in 1989, and it has been spreading west from the eastern U.S. since the early 1980s.

As an aside, if your bees produce honey for human consumption, you should use no treatments without checking with your state entomologist at your land grant college. Your state's agricultural departments may not have cleared certain treatments.

Our wild honeybees get little or no help from humans in combating these mites, and they have disappeared rapidly in the last two years. This is sad! I now have two tube nests for the *Osmia* (orchard bee) in which to place their overwinter stage before the adult dies in summer. They helped save my 1994 crop, as wild honeybees were again scarce to non-existent. Some of our local, small direct-market apple orchards have many sets of these tubes in protected covers to do much of the pollinizing work.

Strangely enough, the African honeybee, which invaded the southern U.S. borders in 1991, has resistance to this mite. This vicious bee at least has that trait to commend it.

Entomologists seem upbeat about the mite problem and point out that the bee may develop defenses—but when? We can no longer ignore our bees or what they accomplish. They need help.

Part II

Descriptions
of 50 Superior Apple Cultivars:
Worldwide Selection and Where
They are Available in the United States.

The descriptions which follow are presented in alphabetical order, only to make it easier for the reader to use the book. Thirty of the cultivars are listed as commercial and are outstanding for homeowners too. Within this group are those that many of us on a worldwide consensus believe are the 10 most important worldwide commercially for the next 15 to 20 years.

---

10 Most Important Commercial Cultivars

Braeburn, Cox's Orange, Elstar, Empire, Fuji, Gala, Golden Delicious, Granny Smith, Jonagold, and Red Delicious.

The 20 cultivars included for the home grower and/or connoisseur either have not been thoroughly tested for commercial production (such as N.Y. #429), or are cultivars that are difficult to make a profit with commercially, although they may be of very high quality. Why some are difficult to grow is outlined in the descriptions.

Following these main cultivars are others usually not pictured. Their names are listed along with short descriptions. Some of these will not be legally available in the U.S. until 1996. After taste testing them, they seemed important to mention. Orders for them should be placed in advance as they will probably be in considerable demand. Examples are Pink Lady, Sundowner, Honeycrisp, Goldrush, etc.

Arlet

Tree:

U.S. Synonym: Swiss Gourmet™, U.S. rights and plant patent #6689 by North American Tree Co. **Origin:** The Switzerland Research Station bred this cultivar from Golden Delicious x Idared, and it was introduced in the early 1980s. A **diploid** of 34 chromosomes. **Size** is moderate to small. Very productive and **annual** if thinned. **Hardy** in USDA zones 5 through 8 and protected zone 4. **Disease susceptibility:** apple scab fungus and powdery mildew. **Sources:** Van Well, Columbia Basin, Burchell and others.

Fruit:

Size is medium to medium-large. **Shape** is round to round-conic. **Flavor** is mild, sweet, and comparable to Gala, with which it will have to compete. **Russets** around the stem and occasionally over much of the apple. **Matures** in late August to mid-September in western Oregon. Warmer areas in the eastern and western U.S. report much earlier ripening, in August. **Storage** life is about two to three months at 32°F. **Stem** is long to very long. Some pre-harvest **dropping** of apples. As of 1995, optimum picking time for storage had not been determined in most areas.

Quality rating: **Very good to Best.**

❖ ❖ ❖ ❖ ❖ ❖ ❖ ❖ ❖ ❖ ❖ ❖ ❖ ❖ ❖ ❖

This new apple has recently joined Elstar (also called Lustre Elstar) and Jonagold as one of the new cultivars grown across northern Europe.

Compared with Elstar, Arlet has a brighter color and seldom has the splotchy orange areas that Elstar exhibits. Also, it does not need the storage time Elstar does to develop its best flavor; however, it is mild and sweet and of fine flavor.

Arlet does not seem to be susceptible to physiological disorders such as bitter-pit, core breakdown, and scald, but it is susceptible to russeting. This cultivar appears to be quite hardy, but there is still much to learn about it as it is one of the newest cultivars to be propagated in the U.S. It does not exhibit heavy spurring.

It begins to bloom at the end of the early bloom period and continues to bloom into the mid-season bloom period.

In Switzerland, Arlet gives slightly better yields than Elstar and has a longer storage life. Arlet is reported to develop an oily skin if stored very long. As of this writing, there are a number of nurseries that have been licensed to grow and sell the trees under the name Swiss Gourmet. It has become a favorite of my family, grown on M.9 rootstock.

It is a dependable and attractive apple without the sharp distinctive flavor of Elstar. Reports from very hot and early areas, such as Arizona, say, it does not keep well as grown there.

Ashmead

✳

Tree:

   Synonym: Ashmead Kernel. A **diploid** of 34 chromosomes of unknown parentage. Tree **size** is moderate. **Origin** was in the Gloucester area of England from a seed planted by Dr. Ashmead around 1700. **Annual** bearer only if thinned. **Winter hardy** in USDA zones 4 through 8. **Disease** resistant to powdery mildew and somewhat resistant to apple scab fungus. **Sources:** many nurseries listed in back of book.

Fruit:

   Size is small to medium. **Skin** is greenish-russet, often a solid gold russet. Sometimes one-half is smooth with no russet. **Stem** is variable, medium to short. **Storage:** three to four months at 32ºF in a refrigerator. **Flavor** is an outstanding rich, high flavor with a balance of sugars and acids. **Pick** mid to late October in Oregon.

Quality rating: **Very good to Best.**

◇ ◇ ◇ ◇ ◇ ◇ ◇ ◇ ◇ ◇ ◇ ◇ ◇ ◇ ◇ ◇

   Of the very old European apples planted in the late 1600s or early 1700s, this one usually wins the most taste tests. Probably only Tydeman's Late Orange or Cox's Orange would begin to vie with it in flavor; however, Cox appeared about 125 years later and Tydeman's Late Orange was crossed by Tydeman in 1930, 230 years later.

   Many years ago I read the descriptive catalogue of Southmeadow Fruit Gardens written by Robert Nitschke (see Bibliography.) Mr. Nitschke is an outstanding connoisseur horticulturist who brought many of these cultivars from overseas and had the patience to send them legally through the disease testing program at the USDA. In his booklet he describes over 200 old European and American cultivars.

   I originally picked eight of them that sounded the most intriguing, and I have never been sorry for ordering them. Ashmead was one. My family and friends can hardly wait for the 1 or 2 bushels we get every year. They regard them as special, as they all like zippy, sprightly apples.

If you like mild, sweetish apples, then Ashmead will probably sting your mouth. It is very high-flavored and a little sharp. The clean aftertaste stays with you awhile. One month of storage improves the flavor.

Apparently 15 years ago many of the nurseries got their grafting wood from an East Coast nursery notorious for virus. Clorotic leaf spot virus by itself usually does no special harm, but when put on an old M.9 with latent virus, sometimes the combination affects the upward shoot growth so it virtually stops and the leaves become very small, almost invisible. It destroyed two trees on M.9 rootstock which I had bought at an estate sale. The source of grafted wood was unknown. My remaining tree came from Southmeadow Fruit Garden and is free of this problem. I mention the above because it is so evident in many of the Ashmead trees I see.

This apple is for the connoisseur and will probably never be grown commercially in the U.S. except by an adventuresome direct marketer. Offer taste tests and I think they could sell for at least $1.50 a pound and be worth every cent of it.

Highly recommended for home yard planting, for fresh eating, in cider and for sauce. On cold-hardy roots, it has survived in zone 3 in Canada for one grower, but USDA zone 4 is probably its northern limit.

Blushing Golden™

✳

Tree:

Trademarked by Stark Brothers Nursery. **Originated** as a chance seedling on the Griffith Farm near Cobden, Illinois, in the 1950s. **Parentage** may be Golden Delicious x Jonathan. A **diploid** of 34 chromosomes. **Size** is medium, but vigorous. **Annual** heavy bearer if thinned. I suggest USDA zones 5 through 7, with trial in lower zone 4. **Disease susceptibility:** powdery mildew, apple scab, and fireblight. **Source:** Stark Brothers Nursery and others.

Fruit:

Size is variable, from small to medium-large. It must be thinned to be **annual**. **Stem** is in a deep cavity and reaches to or slightly above the shoulder. **Texture** is dense and firmer than Golden Delicious. **Russets** somewhat around the stem on at least half the apples. **Pick** in early October in western Oregon. **Storage** life is one of the best, very long in all types of storage. **Physiological problems** are none, except it bruises easily.

Quality rating: **Very good.**

✧ ✧ ✧ ✧ ✧ ✧ ✧ ✧ ✧ ✧ ✧ ✧ ✧ ✧ ✧ ✧

This fine apple is regarded as a new cultivar, even though it was discovered as a chance seedling in the 1950s. It was originally called Griffith apple.

Unless you have personally tried to introduce an all-new seedling cultivar of apple, it is difficult to imagine how long it takes for recognition outside your own area, especially if the apple is yellow and has to compete with Golden Delicious. This is a good example! If you live in the western U.S., you may never have heard of Blushing Golden™. One small grower in the Willamette Valley has offered these apples for sale for a number of years and now has a customer following.

Blushing Golden™ is quite firm and a little tart at picking time, but it is at its prime after one month of storage. A Missouri grower said it was too finicky for his climate. One might want to call this apple an improved Golden Delicious, as it keeps much longer. The flavor is different, almost almond-like, but most who taste it like it very much.

Braeburn

❋

Tree:

   Originated in New Zealand in the late 1940s; introduced in the U.S. in the early 1980s. It is a **diploid** of 34 chromosomes. **Size** is medium after fruiting starts. **Biennial** unless chemically thinned at blossom time. **Hardiness** is unknown but, because of late ripening, I suggest growing in USDA zones 6 through 8, possibly in lower zone 5. **Disease susceptibility:** to apple scab fungus, powdery mildew and fireblight. **Sources:** most commercial nurseries.

Fruit:

   Size of apple varies but it is usually medium, can be medium-small if not thinned. **Shape** is conic to round-conic. **Color** is red blush or red stripes on part of apple, over green. If picked late it is mostly streaky red. **Stem** length is medium to medium-long and usually just above the cavity. **Texture** is firm, slightly coarse and heavy, greenish to white in color. Does not **russet**. **Flavor** is usually balanced and one of the best. **Storage** is long at 32ºF, and up to five months in C.A. **Physiological disorders:** a tendency to bitterpit and internal browning with excess nitrogen use and in hot climates. Five to 10 calcium chloride sprays per summer help eliminate bitterpit.

Quality rating: **Very good to Best.**

◇ ◇ ◇ ◇ ◇ ◇ ◇ ◇ ◇ ◇ ◇ ◇ ◇ ◇ ◇ ◇

   I am convinced this is one of the great new apples of the world. It has a sharp, crisp, high flavor, good appearance and is also a long keeper. It is a heavy apple, high in both sugars and acidity. It does not brown quickly when cut, so it is an acceptable salad apple.

   The cultivar was discovered by O. Moran in Waiwhero, Nelson, New Zealand. Parentage cannot be verified, but may have been from Lady Hamilton seed, open pollinized. New Zealand exported the Braeburn as a test to the larger west coast cities from Vancouver, B.C., to San Diego, California, starting in 1984, at which time our growers had no experience with it.

New Zealand shippers have found ready acceptance for the apple in the U.S. markets because of its excellent flavor and crisp-breaking flesh. This cultivar was introduced with no advertising and no explanation. It appeared in the stores without promotion and sold very well at high prices, further proof that the American public is hungry for better tasting apples.

In New Zealand, Braeburn is picked in late April to May, nearly two months after Gala. In the U.S., it will ripen about with Granny Smith, usually in late October. In 1986 and 1987, my trees of this cultivar were picked on October 27. Fruit from young, heavily fertilized trees are usually of poor quality.

Braeburn probably will not ripen properly north of those areas which cannot ripen other late cultivars such as Stayman, Granny Smith, Sturmer, or Tydeman's Late Orange. One probably can grow Braeburn close to water on the east coast as far north as New Jersey. Testers are doing this now.

My baking and applesauce taste tests with 49 people in 1986 showed Braeburn to be superior for all uses such as fresh dessert, sauce, and baked, as well as for juice.

Carlton Plants sold limited quantities of Braeburn trees starting in 1984 (I got some of my trees several years before this.) Many nurseries are now selling trees in quantity.

In New Zealand, Braeburn mutations of a redder color seem to be of less quality and may be in disfavor. In my opinion, red mutant strains of any apple cultivar usually run downhill in quality.

The Braeburn tree is fairly easy to train with a trellis, and I suggest it be trellised, as mine are. It is grower friendly but, once fruited, the tree grows slowly, with long weeping branches.

Braeburn is highly recommended for commercial or home use in climates with long growing seasons. Only certain areas have outstanding results with Braeburn. It is not for hot areas of California and other warmer areas of the southeast, except perhaps at higher altitudes. In these hot areas, superior management (low nitrogen, many calcium chloride sprays) will help keep it from bitterpit and internal browning—with luck. The Willamette Valley of Oregon seems an absolutely perfect place for low cullage Braeburns. I have one of the oldest trees in the U.S., 14 years old in 1995, and have planted a number of others, all on small rootstocks (M.26, M.9, and M.111/M.9 interstem.) All fruit heavily.

Bramley

*

Tree:

Synonym: Bramley's Seedling. **Originated** as a seedling near Nottinghamshire, England around 1813. **Parentage** is unknown. **Triploid** of 51 chromosomes, so do not rely on it as a pollinizer. Tree is vigorous and with thick, rigid limbs. **Annual** bearer. **Winter hardiness:** I suggest growing it in zones 5 through 6. Buds and blossoms are very frost susceptible. One **source** is Southmeadow Fruit Gardens.

Fruit:

Size of apple is large to very large. **Shape** is flat-round to round and sometimes ribbed. **Skin color** is green unless picked quite ripe, in mid-October here in Oregon. Then it is brownish-orange with little green showing. **Stem** is short and thick, usually within the cavity. **Flavor** is very sharp and acid unless quite ripe. **Drops** badly if overset and quite ripe; pick before then. **Storage** is not over two months in 32°F. I suggest freezing or canning them before Christmas.

Quality rating: **Very good** for cooking.

❖ ❖ ❖ ❖ ❖ ❖ ❖ ❖ ❖ ❖ ❖ ❖ ❖ ❖ ❖ ❖

The British Isles have had and still have a number of good "cookers," but Bramley has that extra acid-apple taste that comes through the sugar in pies and sauce. It is especially valued by the grower, as it is almost immune to apple scab fungus in a climate where scab runs wild. It is also resistant to other diseases such as cedar apple rust, fireblight, and powdery mildew. Another reason for its popularity with the growers is its annual production of big, heavy apples, as apples are sold by the pound.

If one has it planted in a trellis line, as my original tree was, the branches must be trained at once as they become so thick and stiff in two to three years that they tend to break rather than bend. This is a characteristic of a few of the triploids, as are the large leaves.

Setting aside my considerable Welsh and Scottish ancestry, I would probably have been attracted to Bramley anyhow because of its disease resistance, as apple scab is a serious problem here without considerable control in April, May and sometimes early June before the rains stop.

One day at a fruit show I was working on a display that featured what I consider four of the very best apples that originated in England. A very striking lady who had that marvelous English way of speaking said she had grown up in England, and she seemed very interested in the display. I cut off a piece of a ripe Bramley and offered it to her, but she gave me such a horrified look I had visions of the Tower and the Axe. She explained that Bramley was a cooking apple; one ate Cox's Orange fresh. I then showed her the other three English apples in my display, Cox's Orange, Ashmead and Tydeman's Late Orange. She seemed nostalgic and obviously sad there seemed to be no Bramleys available in our stores; she thought little of the cooking characteristics of what apples could be purchased. I sacked up all the Bramleys left, about 12 large apples, and gave them to her, along with the name of a small grower who had about eight trees, so she could again have a segment of "traditional" England in her adopted country. In a small way, British-American relations were probably never better.

If the tree's disease resistance is not important in your climate and you seldom make apple sauce or pie, perhaps you would not plant this cultivar. I will probably always keep a tree of Bramley, if for no other reason than tradition, which I feel helps hold generations together.

Brock

✳

Tree:

Size: medium to medium-large. **Origin:** crossed by the late Professor Russell Bailey of the University of Maine in 1934. **Parentage:** McIntosh x Golden Delicious. It is a **diploid** of 34 chromosomes, and an **annual** bearer if thinned early. **Winter hardiness:** I suggest growing only in USDA zones 4 through upper zone 5 (but also for zones 7-8 of the Willamette Valley.) Quality goes down in warmer areas. **Disease susceptibility:** apple scab and powdery mildew fungus. **Sources** known as of this writing are Roaring Brook Nurseries, Rocky Meadow Nursery and Adams County Nursery.

Fruit:

Size is medium to medium-large. **Shape** is round-conic to mostly conic, similar to Golden Delicious. **Color** is cream-yellow, with pink blush on the sun-side, varying to sometimes considerable reddish-brown over the whole apple. **Stem** length is short to medium, usually within the cavity. **Flavor** is similar to Golden Delicious, but with more aroma and flavor. **Texture** is firm and white. **Storage:** two to three months in cold storage of 30° to 32°F.

Quality rating: **Very good** in eastern zone 4, upper zone 5, and western Oregon zones 7 and 8.

❖ ❖ ❖ ❖ ❖ ❖ ❖ ❖ ❖ ❖ ❖ ❖ ❖ ❖ ❖

This apple is not well known, even in Maine where it was propagated. It first fruited during the time of World War II, and its obscurity seemed sealed until a few years ago. Except for here in the Willamette Valley, it does not do well in the warmer areas of the West and should be avoided there.

I originally saw Brock at a county fair in Eugene, Oregon, in 1983. Shortly after, I traced it to an outstanding small direct-market orchardist, Walter Pope in Lebanon, Oregon. Walter had received scion wood from the University of Maine, and now has two long rows of mature Brock on M.26 rootstock. He grows a number of the better tasting cultivars, but Brock is one of the three most often requested by his customers.

In 1986, Professor Bailey (now deceased) sent me some information which was published in 1966 in *Maine Farm Research*, volume 14, pages 33-36. Below are excerpts from this journal:

Maine selection 7-492 is now released for public trial under the name Brock because it has consistently shown merit during a long period of observation (about 20 years). The name Brock was chosen in part because of interest generated in the new selection by the late Henry Brock, an apple grower in Alfred, Maine. Mr. Brock cooperated with the Maine station in testing several selections. As a result of enthusiastic consumer demand for #7-492 the fruit is sold in his locality under the name Brock. He initiated and supported its release as a named variety. The name is short and there is no record of its previous use as a name for an apple variety.

[...] The original Brock tree made rapid growth and developed into an early bearing tree of desirable shape. Scaffold branches are strong and form wide angles with the trunk. In the nursery, seedling stock budded with Brock has been outstanding in vigor.

[...] Picking season may be spread over a fairly long period because the fruit hangs well, but not too tenaciously to be objectionable. Optimum color and quality develop a few days later than with McIntosh, thus its maturity season more nearly approaches that of the Golden Delicious parent. The fruit is of good size averaging slightly larger than McIntosh and it is similar in shape except that it tends to be somewhat more conic at the calyx end. The calyx is small and well closed. The stem and stem end closely resemble McIntosh. [...] The appearance of fully-matured fruit is appealing and distinctive. The fruit is thick fleshed and has an unusually small core, a most desirable feature for processing and culinary use. [...] Its white to cream color, fine texture and non-oxidizing characteristics make it especially valuable for salads and frozen slices. The flavor of Brock is a mild sub acid and its dessert quality has been rated as equal or better than that of McIntosh in panel taste tests. [...]

[...] Several quality evaluations were made of the Brock variety at different seasons: Newly harvested in November and after storage at 32°F and 90% relative humidity in January, March and April.

Panels of 15 to 31 experienced judges rated the variety for flavor and texture against other seedlings and standard varieties.

[...] Results: In both the measured characteristics, Brock was as good as or better than the 9 different varieties which were used as standards in the 5 tests. Only in April 1964 was it significantly poorer than Golden Delicious.

[...] In conclusion, it is believed that Maine apple selection 7-492 now released to the public under the name Brock deserves a place among new varieties suggested for trial.

After testing Brock for eight years, I would not hesitate to suggest growing it in the Willamette Valley or in zone 4 and upper zone 5. This cultivar seems to ripen in Maine and Oregon at about the same time, October 1 to 10; however, we do not think the apple's keeping qualities as grown here are as good as those reported in Maine. We suggest selling the cold storage fruit in the first two months after picking. Like one of its parents, Golden Delicious, the Brock cultivar has a tendency to bear biennially unless it is well-thinned early in the big bloom year. We have been able to keep the Brock an annual bearer, although some years there are fewer blossoms. It is important not to let it develop a biennial pattern as you may not be able to correct it. It has beauty and very good flavor. It is half McIntosh, and both McIntosh and Brock are at their best in the cooler apple growing areas of zones 4 and 5 and western zone 7. Brock is a north country apple; I suggest not planting it in other areas.

Calville Blanc D'Hiver

✳

Tree:

Synonym: White Winter Calville, Calvite. **Origin** is France in the 1500s. Parentage is unknown. Probably a **diploid** of 34 chromosomes. Tree is quite **vigorous** on M.7A rootstock and a rampant grower. Branches are weepy. **Productivity** is said to be spare in the Midwest, but it is quite heavy here in the Willamette Valley. **Annual** producer as grown here. **Winter hardiness** is not well known, but it does well just north of Detroit, Michigan, in Ontario. I suggest USDA zones 5 and 6, and western Oregon zones 7 and 8. **Disease susceptibility**: to apple scab fungus and very susceptible to powdery mildew fungus. **Sources** are Southmeadow Fruit Gardens and a few other specialty nurseries (see "Nursery addresses").

Fruit:

Size is medium to sometimes medium-large. **Shape** is strange and primitive, with deep ridges on the bottom half of the fruit. **Skin color** varies with sunlight and nitrogen amount. With excess nitrogen, color is quite green to ivory white. With reduced nitrogen, the sun side is orange. It often has reddish-orange spots and **russeting** around the stem. **Stem** length is medium to short, and slender to knobbed. **Flavor** is tart, rich and flavorful. When fully ripe, **texture** begins to soften. **Maturity** is from late September to mid-October in the Willamette Valley; earlier in the southeastern U.S. Thin to one per cluster or short stems will cause push-off of others in the cluster and cause dropping; otherwise, they hang well. Occasionally, it has poor finish from wet spring weather or sprays.

Quality rating: **Very good.**

◇ ◇ ◇ ◇ ◇ ◇ ◇ ◇ ◇ ◇ ◇ ◇ ◇ ◇ ◇ ◇

Calville Blanc d'Hiver is one of the four oldest apples featured in this book. Writers of old mention there were trees growing in the years 1595 to 1598. In the early 1600s it was planted in the Garden of Versailles, and it has been grown extensively into the 20th century by home orchardists and small growers on estates across northern Europe. Nonetheless, it is not included because of its age, but because of its quality.

This apple was brought into Michigan by the French in the 1700s, but apparently was not grown much on the east coast, which was settled primarily by the English and Dutch. In 1903, Beach makes no reference to it at all.

In 1857, E. J. Hooper says in his *Western Fruit Book* that the Calvite was not well liked in humid Cincinnati, Ohio, near the Kentucky border; no reasons were given. However, the former popularity of Calville as a choice dessert apple is indicated by frequent appearances in still life paintings; e.g., Claude Monet's "Apples and Grapes," 1879. Calville is easily recognized in the painting by the deep ridges on the bottom half of the apple.

In reading the many differing reports about it, it is easy to wonder if they are talking about the same apple; I know some of these ancient apples were similar to lesser types. If we are talking of the original, then it appears it is sensitive to climate, soil, nitrogen, or all three. As grown here, I think it is the best pie and sauce apple grown in my orchard over 26 years, and it is one of the best fresh dessert apples. Its vitamin C content is higher than that of other apples and is reputed to be as high or higher than that of an orange of the same weight. In my orchard, growth was only kept under control when on M.9 rootstock. On my current rootstock, M.7A, it is a rampant grower. My best control of its growth has been to bend all three top leaders over and tie them horizontal to, or below, the top wire and let the whole tree and leaders fruit with hanging branches. Robert Nitschke of the Southmeadow Fruit Gardens Nursery tells me it is also difficult to contain in his orchard in the Detroit, Michigan area. I purchased my trees from this nursery many years ago, and at least know we have the same cultivar.

If you like a rich-flavored, multi-purpose apple, this ancient one has stood the test of time, and it should at least be tried in many climates, despite differing reports. It needs early (from bud swell) and thorough spraying for consistent quality.

Coromandel Red

✳

Tree:

   Synonym: Corodel™. Origin: Knotttenbelt Red cultivar (Patent #8460). Of unknown parentage, found in the 1970s. Assigned to Carlton Plants and E. W. Brandt & Sons in the U.S. Tree is of medium vigor and size. Productivity is heavy and tends to biennial bearing. Winter hardiness unknown, but I suggest USDA zones 6 through 8, possibly zone 5. It is a diploid of 34 chromosomes. Disease susceptibility: apple scab fungus. Sources: Brandt's Fruit Trees, Inc. or Carlton Plants.

Fruit:

   Size is medium to medium-large. Shape is round to round-conic. Skin color is dark pinkish-red over yellow. Stem is medium long to very long. Flavor is sweet-tart. Texture is firm, dense, with greenish to white-yellow flesh. Storage life needs more testing, but it does appear to be more than two or three months. Matures in late October in western Oregon and western Washington, west of the Cascade Mountains. Physiological disorders are unknown, but so far none have been noted.

Quality rating: Very good.

❖ ❖ ❖ ❖ ❖ ❖ ❖ ❖ ❖ ❖ ❖ ❖ ❖ ❖ ❖

   This is a new apple from the Coromandel Peninsula in the Hawke Bay area of New Zealand. Because of its fine flavor and very handsome appearance, it warrants inclusion in this book.

   My own trees have fruited four times, but I have also tasted apples shipped in from New Zealand some years ago; the taste was way above average. It may be another sleeper that should be tested over a wider area.

   This cultivar reminds me of the original Red Delicious in flavor but it is more attractive, with distinct dots on a dark pinkish-red skin. The apples hang well and do not pre-harvest drop. On December 1, one of my trees had nine apples still on the tree (M.26 root), so it seems they do not even drop their fruit past normal picking time.

Americans seem to like good foreign apples, and here is a rather different one with an intriguing-sounding name to take to your local fruit show. Coromandel Red is a connoisseur apple that may have direct market commercial potential.

Cortland

✻

Tree:

Origin: New York State Experiment Station, Geneva, N.Y., crossed in 1898, introduced in 1915. **Parentage:** McIntosh x Ben Davis. **Diploid** with 34 chromosomes. Tree is moderate in vigor and **size**. **Annual** in bearing habit. **Winter hardiness:** similar to McIntosh. I suggest USDA zones 4 through 6, and some of lower zone 3 protected by water. **Disease susceptibility:** very little, except very prone to powdery mildew fungus in mild winter climates such as here in the Willamette Valley. **Sources:** Most commercial nurseries.

Fruit:

Size is medium to medium-large. **Shape** is round-oblate, quite flat. **Skin color** is red and red stripes over yellow-green background. Solid pink-red strains are available. **Stem** length is usually above the cavity or level with it; some stems are knobbed on the end. **Flavor** is rather sweet-tart, but not intense. Pleasant taste with less aroma than McIntosh. **Texture** is coarse, juicy, with pure white flesh which becomes rather soft when ripe. **Matures** in late September to early October. Stores at 32°F for two months. **Russeting:** seldom.

Quality rating: **Good to Very good.**

◇ ◇ ◇ ◇ ◇ ◇ ◇ ◇ ◇ ◇ ◇ ◇ ◇ ◇ ◇

This cultivar was developed by Professor S. A. Beach in 1898, and it should be noted that Professor Beach was also the primary author of two excellent volumes titled *The Apples of New York*. These books are now rare and, when they can be found for sale, quite expensive. Because the Cortland apple cultivar was not introduced until 1915, it was not described in his books.

The tree is not large on size-controlling rootstock, but it is quite large on seedling rootstock, which I do not recommend. It is rather upright-spreading and produces fruit spurs freely. The cultivar is a pollen-viable diploid, so it can be used for cross-pollination. It does not pre-harvest drop like McIntosh.

I believe the Cortland to be easier to grow than McIntosh, with a flavor comparable to McIntosh but not as perfumey or quite as high-flavored. It is only a fair processing apple. It is one of the few really good salad apples because, after slicing, the white flesh does not oxidize or turn brown for hours.

Cortland was in 10th place in the top 15 U.S. commercial apples in 1988, with approximately 3.5 million 42-pound boxes produced annually. A fine, no-nonsense commercial apple for the north country. Like the McIntosh, the best quality Cortland is in USDA zone 4.

Cox's Orange

✳

Tree:

Synonym: Cox's Orange Pippin. Origin: by Richard Cox of Coln Lawn in Buckinghamshire, England, around 1825. Parentage: open-pollinized seed of Ribston. A diploid of 34 chromosomes and completely self-sterile, but it pollinizes other apple cultivars. Tree is moderate in vigor and size. Productivity is average to low. Annual in production. Hardiness is unknown, but I suggest the cooler parts of USDA zones 5 through 7. Sources are a number of small specialty nurseries; those I'm aware of are listed in the back of this book. For over 100 years, Cox's Orange has been the most popular dessert apple in England and parts of northern Germany.

Fruit:

Size is medium to sometimes medium-large. Shape is round to round-oblate (flat). Skin color varies from bright orange-red to reddish-brown, sometimes with russet over the top half of the apple. Stem length is usually medium, just above cavity. Flavor is a rich, nutty, outstanding blend of sugars and acids. Matures in early- to mid-September in most northern zones. Stores for two to three months at 32°F.

Quality rating: Very good to Best.

◇ ◇ ◇ ◇ ◇ ◇ ◇ ◇ ◇ ◇ ◇ ◇ ◇ ◇ ◇ ◇ ◇ ◇

After reading the rather short account in Professor S. A. Beach's *The Apples of New York*, I found it strange he recommended this cultivar for the home orchard, as he said it was undesirable for commercial planting but gave no reasons for this advice. So, many years ago I set out to discover the reasons by planting a Cox's Orange on M.26 rootstock in heavy clay. Unintentionally the tree received rather poor care, including inadequate watering. M.26 rootstock does not tolerate insufficient water. My first two crops resulted in 85 percent of the apples cracking to the core and rotting. That was my first acquaintance with the frustrating genetics of Cox's Orange, frustrating especially in the first two or three years of fruiting on juvenile trees. Correct cultural conditions will compensate or control some of these problems, and spraying with the hormone gibberellin GA4-7 at blossom time and two weeks later should help control cracking. Three calcium chloride sprays also help, and a little extra fertilizer and water seem advisable too.

There are now a number of different strains of Cox's Orange, one of the better ones being a bud mutation named Queen Cox. New Zealand has developed some new strains from irradiation which have not yet been released for testing. New Zealand strains are reported to have striped color, minimal or no cracking, and flavor similar to the original Cox. How well they will perform across North America has yet to be determined. I prefer some first generation crosses of Cox's Orange, especially Tydeman's Late Orange and Karmine™, as I think they are much easier to grow and also have outstanding quality.

Where nights are usually cool in September, as they are here in the Willamette Valley, Cox's Orange frequently colors a beautiful orange-red on one-quarter to three-quarters of the apple. There may be a light gray-brown russet around the stem and sometimes on one-half of the apple; however, once it is tasted the taster cannot remember the color of the apple. It simply does not matter anymore! Yes, it tastes that good, and it always ranks at or near the top in taste tests. In the pre-harvest hot areas like eastern Washington, the quality is usually poor. Cox likes a mild, perhaps even rainy, summer such as we find in England, the Benelux countries, and western Oregon, west of the Cascade Mountains.

No matter that it is susceptible to scab and mildew, you must simply give it a little extra tender loving care. If these trees are watered and limed properly, the fruit usually does not crack after the first crop or two.

England, in many areas, has limestone soil with a pH of 7.5 to 9. Perhaps that is one reason Cox does well there. Many of its first generation crosses, especially Holstein, often crack the first year or two of production. My advice is to have patience with Cox, and let it crop for a few years; make certain to apply limestone steadily to get the pH at least to 6.5.

In England and northern Europe, Cox production per acre after the sixth year seldom exceeds 500 to 600 42-pound boxes. The late Dr. McKenzie of New Zealand advised me that Cox production with some of the new clones there occasionally reached 1200 42-pound boxes per acre, which is high, and partly due to a more favorable climate. Due to low production and its popularity in England, Cox sells for more money per pound. Commercial growers in North America tend to shun Cox as somewhat unmanageable.

Delicious (Red)

✼

Tree:

   Synonyms: the original name (before 1894) was Hawkeye; then Delicious; now called Red Delicious. **Origin:** the farm of Jesse Hiatt near the village of Peru, Iowa. **Parentage** is unknown, but Hiatt thought it was from a seed of the nearby Yellow Bellflower. It has 34 chromosomes and is a viable pollen **diploid:** $2 \times 17 = 34$. Tree is vigorous and healthy and of medium **size.** Rather slow to start bearing, one to three years later than most commonly grown apples. **Annual** bearer if properly thinned. **Winter hardiness** to USDA lower zone 5 and through zone 8, zones 6 through 8 in Washington State. **Disease susceptibility:** apple scab fungus on both fruit and leaves. **Resistant** to fireblight and powdery mildew. **Blossom season:** mid- to late-bloom. **Sources:** most commercial nurseries. There are many strains of solid or striped red, with differences usually slight.

Fruit:

   Size varies from small to large. **Shape** is unusual. Many strains have flaring points on the calyx end intensified by cool spring weather and the use of promalin, a plant hormone. **Color:** the original Hawkeye was green with red stripes and blushes. Strains since 1922 are solid red, often with dark red stripes before ripe. **Stem** length is medium long to long on most strains. **Flavor** is sweet and rather bland. Liked by many if fruit is crisp. Flesh **texture** is firm and **color** should be white to cream white, not green and starchy from early picking. I have never seen it russet. In eastern Washington **picking** is usually mid- to late-September. **Storage life** is long in C.A., up to 10 months; three to five months in common storage. Its **physiological disorder** is primarily water core brought on by poor orchard management, late picking, too hot weather, and perhaps some genetic susceptibility.

Quality rating: Original Hawkeye/Delicious **Very good**; modern red strains, **Good.**

⬦ ⬦ ⬦ ⬦ ⬦ ⬦ ⬦ ⬦ ⬦ ⬦ ⬦ ⬦ ⬦ ⬦ ⬦

   This is the most commercially grown apple cultivar in the world, and it was nearly destroyed and lost many times before it was known to the general public.

Only Northern Spy had such a close call before being fruited, but that happened only once.

Some time between 1868 and 1872, a young apple tree sprouted from seed in farmer Jesse Hiatt's orchard near the village of Peru, Iowa. The village is near the town of Winterset, a few miles southwest of Des Moines in the south-central part of Iowa, which is the upper part of USDA zone 5.

Mr. Hiatt is supposed to have cut down the young Delicious tree at least once because it was not in a row with his other trees. Another reason he may have wanted to remove the tree is because seedlings normally produce a very low percentage of useful apples.

His other trees were all grafted with many cultivars, as that practice was common by then and it is known that Mr. Hiatt was an adequate grafter. The tree continued to grow back up and, being a Quaker, his religious beliefs perhaps helped fuel his compassion for the struggling young tree, and he started to care for it. The Delicious tree was growing near an old variety called Yellow Bellflower. He liked the Yellow Bellflower apple, and if Delicious was a seed from that, perhaps the new tree would become another variety worth keeping.

The Yellow Bellflower is shaped somewhat like the Delicious apple, but it has a very different color and taste; however, one always has this genetic diversity in apples to deal with. It seems best to go with Jesse Hiatt's reasonable assumption that the Delicious seed came from the nearby Yellow Bellflower. (I think it more nearly resembles the old Black Gilliflower.)

He cared for the tree for 10 years. Finally, it had a blossom cluster, but only one apple came from this cluster. It had such an excellent taste he told his wife that this was the best apple in the world and was to be called Hawkeye.

Then the second close call occurred that almost destroyed this seedling. Because of winter sunscald, the tree became scarred on the trunk and much of the bark peeled off on one side. The tree was in danger of dying, so a heavy cover was put on the trunk in wintertime for protection from quick temperature changes brought on by daytime sun and nighttime cold.

As the years went by, Jesse was unable to interest anyone in his Hawkeye apple. Some people made fun of its strange shape because it had long, ridged points on the calyx (basin) end. Most tasty apples in those days were round or conical. Hiatt thought that sending his apple to the county fairs in Iowa would help it gain some recognition, but the Hawkeye won no ribbons. All of the prizes at that time were being won by Jonathan, Baldwin, Northern Spy, Wealthy and others.

Perhaps another man would have given up, but Mr. Hiatt combined stubbornness with zeal, an equation for success. In 1893, he sent a few Hawkeye apples to a fruit show held by Stark Brothers Nursery in Louisiana, Missouri, about 70 miles north of St. Louis on the Mississippi River. When the judges were through with their investigations, they gave the Hawkeye apple a first prize. Then C. M. Stark tasted one, and immediately gave it a name he had written down and kept for a long time in a notebook. The name was meant specifically for a superior, future new fruit and Hawkeye apple was definitely in that category. He named it Delicious!

So, was all of Mr. Hiatt's work finally to be rewarded? Not quite! The variety had another close call. After the first prize award, Stark felt it was paramount that the owner be contacted at once but, with all the activity at the apple show, Mr. Hiatt's name and address had been lost. Jesse did not even know his apple had won a prize.

Also unknown to him, he and his Hawkeye had probably the most powerful ally they could have: C. M. Stark of the famous Stark Brothers Nursery. There is evidence that Stark put his fruit show on again the very next year (1894) with the hope that the Delicious apple would be entered so he could obtain the name and address of the man who had sent it. C. M. Stark did not know that Jesse Hiatt was almost 70 years old and time was running out, as about 24 years had passed since the little tree had grown up near the old Yellow Bellflower.

In 1894, the team combination was now a strong persistent apple tree, a stubborn Quaker farmer, and an excited and famous nurseryman. The fruit show opened that next fall and everyone wondered if the new, strange-shaped Delicious apple would be entered again. Transportation being what it was at that time they were fortunate the apples arrived at all, but a big barrel of Hawkeye apples was delivered—with the sender's name and address attached. The fruit, streaked with red over green and yellow, was ripe and had an unusual, piquant flavor.

Stark immediately wrote to Jesse Hiatt. In his reply, Mr. Hiatt said that if Stark Brothers Nursery did not think the Hawkeye was better than other varieties, they could have the rights to propagate it for nothing. Stark Brothers agreed that it seemed a superior variety and paid Mr. Hiatt well.

C. M. Stark then found that the Delicious tree had many good characteristics. It had strong branches that formed wide angles and held heavy loads. The apples held well on the tree. Also, Delicious was a rare annual bearer that did not need much pruning. Perhaps most important was the fact that the cultivar had a considerable resistance to the serious bacterial disease called fireblight. The apple had fairly good color for its day and kept well if not picked too ripe. The flavor was somewhat bland but had an attractive indefinable quality and aroma. It was also reasonably hardy for southern Iowa and had withstood drought and hard winters.

It was discovered in later years that the Delicious grew well in many types of soil and in most climates favorable to apple growing. It was, and is, grown in many countries. Only a few apple cultivars have qualities such as these. Like the old Ben Davis, Delicious is a grower's dream provided the apple scab fungus (*Venturia enequalis*) is controlled early. In eastern Washington's dry climate, they seldom need to spray for apple scab. Delicious also stands their hot summer climate quite well. This area now has about 110,000 acres of Delicious in Washington alone.

But the chronicle of the spunky original Delicious tree is not over. On November 11, 1940, there was the storm of storms—the infamous Armistice Day Storm. It was unusually severe for two reasons: it came so early in the fall, and with an unheard-of temperature drop from about 60°F on November 11 to minus 5° by noon November 12, a 65°F drop in one and a half days. I went through this "northern screamer" at our home in Northfield, Minnesota. As the storm descended on Minnesota, Iowa, and other surrounding states, it gave me a strong feeling of apprehension as I walked to and from classes at St. Olaf College. By 3 p.m., I was looking at snowflakes blowing horizontally instead of falling vertically, while the temperature dropped 3° to 7°F per hour. I remember this storm in hour by hour detail, perhaps because of the alarming temperature drop with warnings on the radio too late, or that so many trees were killed, or because acquaintances who were duck-hunting died that night of hypothermia—or probably due to a combination of these things.

At our farm four miles away from our Northfield, Minnesota home, none of our Haralson, Wealthy, Duchess, Whitney Crab and other crabs were severely damaged. They were hardy types that defoliate earlier, and a west and northern conifer windbreak also helped.

Farmers in Iowa, another zone south, had planted many Delicious and Northern Spy trees. By late spring, Iowa extension and research people told me they found no Delicious apple trees alive in Iowa and not many Northern Spy. Thousands of apple trees in Minnesota, Wisconsin, Iowa and other parts of the upper Midwest were so damaged that day they were dead by spring. Two hundred and eighty miles south of our farm, the world famous original Hawkeye/Delicious tree that was the mother to millions of grafted Delicious trees was dying. By the spring of 1941, it was mourned by horticulturists and growers everywhere.

You should be aware that this apple had already become a "mortgage lifter" in many parts of the nation. Delicious apples had put money in the pockets of often beleaguered orchardists from coast to coast. All the publicity about the "dead" original Delicious tree—obituaries appeared in newspapers and on the radio—occurred because many growers cared about the tree. After all, it was only just past middle age, approximately 70 years old. Since there were countless seedling rooted apple trees in the U.S. and Europe that had reached 150 years of age or more, Delicious should have had many years left.

The pronouncement of the tree's death was premature! Only the *top* of the original tree had died. Just as when Mr. Hiatt had cut the tree down, the roots again displayed their remarkable vitality. After that awful storm of 1940, two shoots from the roots formed two new trees about 14 feet apart and, in a few years, these bore annual crops. They were still giving apples decades later. The original Hawkeye had an aroma the red sports do not have, and I think the flavor is definitely better than the red strains. I grew up eating the old Hawkeye and have encouraged a movement to bring it back. Why not, at least for home use? Stark Bros. started producing it again in 1993.

Present day Red Delicious

At the turn of the 20th century, horticulturists, growers and customers were enthusiastic about the tree, its fruit, its unusual shape, flavor and good keeping qualities. Today there is a different perception of Red Delicious. Many people have mixed feelings about the red strains and their over-supply sometimes, when the wide diversity of other apple cultivars could be offered to the consumer in greater quantity.

I have been involved with taste tests of many apple varieties since the late 1960s. Of the more than 11,000 people who participated in these tests, about 20 percent say they never buy any Red Delicious apples but do buy other apples. Those who do buy them often say that the apple is mushy and tasteless. So what happened between 1900 and 1990 that brought the apple from having universal appeal to some public dissatisfaction?

Over the history of this apple, many changes have occurred. The new dark red spur strains tend to runt out on some Malling rootstocks and are fed too much nitrogen to keep them vigorous; they are often picked too early, stored a long time in C.A., and handled poorly in stores.

Obviously, many people do eat Red Delicious. Just out of storage the apples are fine. But in July and August the produce managers of stores use much of store refrigeration for fresh vegetables, which are more perishable, and stores often do not have enough cold storage space for the apples as well. In addition, some store managers think the Red Delicious apples will keep when stored at room temperature for 2 to 3 weeks. I know they usually will not; if they have already been stored many months, they should be kept under refrigeration. Store produce managers should be helped by education on this issue. I have talked to many store produce managers and have learned that the entire system of growers selling to packers, packers selling to chain stores or brokers, and finally, stores selling to consumers, often breaks down at certain points, especially in the retail stores.

It appears that eastern Washington does need to have much of the system of distribution it uses because of low population and the over 100 million boxes of all apple cultivars produced yearly, of which 110,000 acres are Red Delicious alone. There are also another 50,000 acres of other apple cultivars according to a 1986 Washington State University survey—160,000 acres of apples total, on which about 50 percent of the nation's apples are grown. Eastern Washington growers need to better protect this enormous potential income.

Red Delicious will no doubt be with us a long time, but I think its quality can and should be improved by certain changes, from grower to consumer.

Jesse Hiatt died in 1898, without ever knowing how economically important his Hawkeye would become.

Elstar

✳

Tree:

Synonym: Lustre Elstar in the U.S. under plant patent #6450 by Carlton Plants, which also has licensed Stark Brothers and others to sell trees. **Origin:** The Netherlands, about 1955. **Parentage** is Golden Delicious x Ingrid Marie. Since Ingrid Marie is half Cox's Orange, I could say Elstar has impeccable bloodlines. **Diploid** with 34 chromosomes. Tree is of medium vigor and medium **size**, with a tendency to **biennial bearing**, unless thinned well. **Winter hardiness** is unknown, but I suggest growing in USDA zones 5 and 6, with trial in protected zone 4. **Disease susceptibility:** apple scab fungus, fireblight, and powdery mildew fungus. **Sources** are Carlton Plants; Brandt's Fruit Trees, Inc.; Stark Bros.; Van Well and others.

Fruit:

Fruit **size** is medium to occasionally medium-large. **Shape** is round to conic. **Skin color** is yellow with mottled orange in young trees; more orange-red on older trees. **Stem** length is medium to quite long and usually above the cavity. **Flavor** is outstanding, but quite sharp and tart if picked early enough for storage of two to three months. **Texture** is crisp and fine, with cream-white flesh. **Russets** occasionally around the stem. **Matures** from August 25 to September 10 in Washington and Oregon.

Quality rating: **Very good to Best** in cooler pre-harvest climates.

❖ ❖ ❖ ❖ ❖ ❖ ❖ ❖ ❖ ❖ ❖ ❖ ❖ ❖ ❖ ❖

This is a relatively new apple from The Netherlands. As of this writing, Elstar has been heavily planted in northern Europe for at least 20 years. It is a tart, mid-season apple of outstanding flavor. The reports of high quality from Europe convinced me that this cultivar should be tried in this country, so some years ago I suggested to the then Carlton Nursery that they acquire the rights for Elstar. They did, and good quantities of trees are now for sale.

The fruit ripens around September 5 to 10 here in the Willamette Valley of Oregon, and as early as August 20 to August 30 in the warmer Yakima Valley of

Washington. Expect it to be tart in flavor, but it mellows in storage. I really like its tart-sweet flavor, and it wins many taste tests.

This cultivar will produce better quality apples in the cooler areas of the U.S., and in portions of northeastern Canada such as Nova Scotia. It has become one of my top four all-purpose apples.

Like other Golden Delicious crosses, it seems sensitive to nitrogen at levels above 2.00, leaf analysis in early August. Do not apply nitrogen after April, or the Elstar trees may not lose some of their leaves in the fall or become fully dormant; at least this is the situation in our mild climate. Since MM.106 rootstock tends to induce late dormancy, I would not use it for this apple if you have hard freezes in late October or early November. Excessive nitrogen use also shortens the storage life of the apple.

If picked early enough, Elstar apples keep quite well for two to three months. In The Netherlands, it is picked later than we pick ours because of their cool, cloudy summers with very few 85°F days.

In personal conversation with Tom Vorbeck, one of the nation's most knowledgeable apple growers and owner of "Apple Source," he said Elstar usually does not do well in hot, humid southern Illinois. Warm nights probably are a limiting factor. High pre-harvest temperatures almost always lower quality in August or early September ripening apples.

Highly recommended for fresh dessert, sauce, pie, and microwave baking.

Empire

✳

Tree:

Synonym: New York #45500-5 before it was named. **Parentage** is McIntosh x Red Delicious. It is a **diploid** of 34 chromosomes. Tree **size** is medium to small. **Multiple spurs. Annual** bearer if thinned. **Hardy** from zone 4 through 8. **Disease susceptibility:** very susceptible to apple scab fungus in my climate. It appears to be somewhat resistant to fireblight and powdery mildew.

Fruit:

Size is usually medium, small if not thinned. **Shape** is round to round-conic. **Skin color** is dark red on the sun side to yellow on the underside. **Stem** length is medium, just beyond the cavity, longer than Liberty. **Flavor** is sweet. **Texture** is crisp and breaking. **Flesh** snaps off in chunks. Seldom, if ever, **russets. Ripens** in mid-September in the north. **Stores** two to three months in cold storage at 32°F. At 45°F it may be soft and mushy before Christmas. Best in C.A. storage.

Quality rating: **Very good.**

◇ ◇ ◇ ◇ ◇ ◇ ◇ ◇ ◇ ◇ ◇ ◇ ◇ ◇ ◇ ◇

In 1945, Dr. Roger Way at the New York Geneva Station produced Empire. It was first fruited in 1954, tested as N.Y. #45500-5 and introduced in 1966. McIntosh has been used as one parent in dozens of crosses in the hope of producing a better apple than either parent. Empire is a grower's apple because it is relatively easy to produce an annual crop of a crisp, attractive fruit that keeps fairly well, tastes above average, and does not tend to drop much of its crop before ripening, as McIntosh does. I think the perfumed McIntosh produces top quality fruit only where September nights are in the 35° to 50°F range. Such temperatures foster high color and flavor. McIntosh's offspring Empire is not so finicky, as it can produce good quality fruit from the southern Appalachians of northern Georgia to southern Ontario near the Great Lakes. In some microclimates it does not color well, but that is the exception, not the rule.

This apple is best for fresh dessert, but it is acceptable for pies and cider mix. Empire seems only fair for sauce and drying. Apple identifiers usually find it difficult to distinguish the fruit of Empire from Liberty. The resemblance is remarkably close, with only slight differences in stem length and the crowns on the bottom of the apple. Empire is sweeter and Liberty usually has whiter flesh.

The tree has good characteristics: wide branch angles, abundant fruit spurs. However, Empire may do better on a rootstock such as M.7A, instead of M.26, as it stimulates a more vigorous tree. Multiple spurring is dwarfing, and it needs a more vigorous root.

Empire was being heavily planted in eastern North America. It is an early-season bloomer in mild areas such as California and Oregon but it is a mid-season bloomer in Ontario, Canada. Because of its small size and rather short storage time in regular storage, its long term commercial prospects are unclear.

Erwin Baur

�֍

Tree:

Origin: Muncheberg, Germany, about 1928, and reported as a seedling of Duchess of Oldenburg, open pollinized. A **diploid** of 34 chromosomes. Tree **size** is moderate to small. **Annual** bearer if thinned, and very heavy producer. I suggest USDA zones 4 through 8. **Disease susceptibility:** anthracnose. **Resistant** to apple scab fungus. **Blooms** very early. **Source:** Southmeadow Fruit Gardens. I am not aware of other sources, although the eastern U.S. may have other nurseries propagating it.

Fruit:

Fruit is medium in **size** and round to round-conic in **shape**. **Stem** length is medium to long, and slender. **Color** is bright scarlet red, with yellow background. **Flavor** is tart. Flesh **texture** is crisp and dense, and mellows after a month in storage. Makes applesauce that rivals that of Gravenstein. No **physiological problems** known.

Quality rating: **Good to Very good.**

◇ ◇ ◇ ◇ ◇ ◇ ◇ ◇ ◇ ◇ ◇ ◇ ◇ ◇ ◇ ◇

This is a beautiful apple, named after the founder of the Institute of Plant Breeding in Muncheberg, Germany. Its resemblance to the Duchess of Oldenburg cultivar is hard to detect, except for a very sharp flavor. It is one of the most highly flavored apples I have ever grown. Many have mistakenly heard that its high flavor is just acidity, which may be the main reason why it is not better known. If one waits 10 seconds after biting into it, there is a fine mouth-cleaning flavor. I prefer its apple sauce to that of Bramley. It is also a much easier tree to manage. Erwin Baur is excellent in a cider mix, giving needed acidity and flavor.

If this apple really is half Duchess of Oldenburg, then it should grow fairly far north, as far as such a late-ripening apple can grow, such as perhaps southeastern Minnesota.

Meager information is available from others, but I have detected no bad tendencies. It should be grown in areas with a medium-long growing season. In the nine years it has fruited in my orchard, it is an annual bearer if thinned. If picked before approximately October 1 to 15, it is too acidic for most tastes. Erwin Baur stays well on the tree, and in the north would be classed as a winter cultivar.

In southern Ontario, just above Detroit, Michigan, this winter cultivar has won taste tests over most other late apples, but some of my friends do not like its sharp flavor.

Idared blooms as early as Erwin Baur and also ripens in October, and they have good bloom overlap (usually early bloomers ripen in August or September.) A rare cultivar; more of us should discover it, especially those of us who like a sharp, tart flavor.

Freyberg

*

Tree:

Origin: crossed by plant breeder J. H. Kidd of New Zealand, around 1934. **Parentage:** Golden Delicious x Cox's Orange. **Diploid** of 34 chromosomes. Tree is of moderate vigor, but in **size** it is rather small and weak. **Annual** bearer if thinned, but it is not a heavy producer. **Winter hardiness** is unknown to me because very few people in the U.S. grow this cultivar. I suggest growing it in USDA zones 5 through 7, with trial in zone 8. **Disease susceptibility:** apple scab fungus. **Sources:** Southmeadow Fruit Gardens, Rocky Meadow Nursery, and Raintree Nursery.

Fruit:

Size is medium to sometimes small. **Shape** is round-oblate, rather flat. **Skin color** is light green to pale yellow; occasionally has a light grey **russet**. **Stem** is medium to rather long, usually the latter. **Flavor** is an outstanding sweet to very sweet with a cocktail of flavors. **Texture** is fine and juicy, and **flesh** color is cream-white to white. **Matures** here in early October.

Quality rating: **Very good to Best.**

❖ ❖ ❖ ❖ ❖ ❖ ❖ ❖ ❖ ❖ ❖ ❖ ❖ ❖ ❖ ❖

This is a very high quality apple developed by J. H. Kidd of Greytown, Wairarapa, New Zealand, the same horticulturist who brought us Gala and Kidd's Orange Red.

The Freyberg is one of the sweetest apples I have eaten. Many specimens exhibit a blushed pinkish-brown on the sun side. Ordinary appearance is the downfall of many good-flavored yellow or green apples, and Freyberg is seldom distinctly yellow, especially if given excess nitrogen. Virtually none of the Golden Delicious crosses, including Freyberg, Jonagold, Elstar and Gala, should contain more than 1.7 to 2.0 percent nitrogen by leaf analysis in early August, or they may not color well or even ripen properly.

The apples keep only until about Christmas. The Freyberg fruit is not found for sale in U.S. markets and is primarily for the connoisseur.

Freyberg seems to tolerate summer heat and warm nights just above the Mason-Dixon line (Southern Illinois, Southern Indiana, etc.), as well as our cooler climate.

Winston Churchill, in his book *The Grand Alliance*, mentions he entertained a young New Zealand officer, Bernard Freyberg (1898-1963), at a country estate in the early 1920s. Churchill reported that Freyberg had 27 scars from shrapnel and bullet wounds, many of which he received at Gallipoli in 1916 where he earned the Victoria Cross. Churchill said he was only one of two men he ever knew that had been "shot to pieces" and seemed to fully recover, both physically and psychologically. In World War II, Freyberg became a general and led the 2nd New Zealand force in battles in Greece, North Africa, and Italy, during which time he received three more scars. As Sir (General) Bernard Freyberg, he was governor-general of New Zealand from 1946 to 1962.

I can't believe our New Zealand friends would name a tasteless apple after so illustrious a man, and they did not. It is a remarkable apple of many flavors, and aptly named after General Freyberg.

A true domestic cultivar that you must grow yourself.

Fuji

✳

Tree:

Synonym: was Tohoku #7. **Parentage:** a cross of Red Delicious x Ralls Janet. A mid- to late-blooming **diploid** of 34 chromosomes. Moderate **vigor**. **Medium-sized** tree. **Biennial bearing** habit unless carefully pruned and thinned early. I suggest USDA zones 6 through 9 until tested more in zone 5. **Disease susceptibility:** fireblight and Alternaria spot. Somewhat resistant to apple scab fungus. **Sources** are most commercial nurseries, at least in those areas where it will ripen.

Fruit:

Size is usually medium, sometimes small or medium-large. **Shape** is round to conic. One of the most variable of apples in **color**, it varies even in the same field, especially on the redder strains. The original strain usually has light pink or purple stripes over a green background. In hotter areas it often is a rather ugly yellow-brown. **Stem** length is medium-long to long. **Flavor**, when sugar percentage is high enough, is very sweet and appealing; otherwise, it is quite flat and unappealing. **Russets** and rain-cracks in many climates. **Matures** from October 20 to November 11 in the northwestern U.S. **Storage life** is long, one of the best. **Physiological disorders** are water core, cracking and storage scald.

Quality rating: **Very good** if sugar content is high enough; otherwise, not over **Good**.

◆ ◆ ◆ ◆ ◆ ◆ ◆ ◆ ◆ ◆ ◆ ◆ ◆ ◆ ◆ ◆

Most of the testing so far is being done in the west in zones 6 to 8. Parts of zone 9 seem too hot, resulting in sunburn, poor color, russeting and cracking in 30 to 40 percent of the crop.

There is a description of one parent, Red Delicious, in this book, but something should be mentioned about its other parent, Ralls Janet. This cultivar was first noticed growing on the farm of Caleb Ralls in Amherst County, Virginia, in the 1700s. Its origin is not known for sure and has been steeped in controversy for 200 years. It became a favorite apple through much of the south and across the. Mason-Dixon line, growing in many of the same regions the Ben Davis did later. Ralls Janet can be colored yellow or green, with a light overlay of faint pink and gray. This apple has a sweet, aromatic, pleasant flavor and the tree blooms late. Beach gave it the high rating of "very good" for fresh dessert.

Japan's two centuries of seclusion ended with the arrival of Commodore Matthew Perry and his four naval ships in 1856. Within a year of this initial contact, a treaty of friendship was signed. Not too many years later, apples were shipped to the Japanese people. This resulted in their planting some of our established cultivars, one of which was Ralls Janet. The Japanese like its sweet flavor and late blooming characteristic.

The cross breeding resulting in Fuji is reported to have been done in 1939 by H. Niitsu at Fujisaki in Aomori. In 1958 the new apple was named Tohoku #7 after the Tohoku Research Station. In the 1960s the Japanese thought so much of this apple they renamed it after their symbol of purity, loftiness, constancy and majestic beauty, the nation's highest mountain, Fuji.

Currently Fuji accounts for about 45 percent of the total apple production of Japan. This is remarkable when one considers they have more cross breeding of apples than most other countries. (The Japanese have about 15 to 25 new apples currently being tested, some of which are reported to be above average.)

In areas where Fuji does not mature properly, the flesh may be a bit woody, hard and tasteless. Keep in mind that in areas where no growing or testing of a cultivar has been done, proposing an area where it might grow well is speculation. A lot of money should not be invested in any area until after testing the cultivar. Fuji is site-selective for climate. One must test and, yes, it may take years.

My Fuji trees have excellent shape, with strong wide angle branches. One is a strain called B.C. #2 (Mori Ho-Fu #2). In the first two years these trees fruited, they consistently produced apples of a reddish-brown, unattractive color, but it is important for the reader to know that the Fuji cultivar does not appear to produce good colored, high quality fruit on young trees. The Fuji tree usually must fruit at least three to six years, unlike Gala or Braeburn, before the high quality of the apple becomes apparent. Fuji appears to me to be extremely nitrogen-sensitive for the first three or four years.

This cultivar has a genetic tendency to water core, aggravated by late picking. The Japanese pick Fuji late in order to obtain better skin color and some water core. The water core has a syrupy, sweet flavor which the Japanese people apparently like. Other physiological disorders are russeting, cracking and apple scalding in storage (brown areas developing on the skin), especially when it has been grown in very hot areas.

The tree appears to be quite hardy after going dormant in winter. It is difficult to know precisely how hardy, because the apple matures so late that one cannot ripen Fuji very far north. I pick mine in early November and, in my opinion, the flavor of the original strain is better than some of the new color strains I have tested.

To be a tasty apple, Fuji must get a high sugar (carbohydrate) content. In 1987 my trees had a huge overset of apples as I did not thin them enough. There were not sufficient leaves to form the necessary sugars, and the apples were very tasteless. In 1989 and 1990 I corrected this and got a better flavor. One tree on M.26 rootstock is now 14 years old, and the additional tree age also helped the flavor. A minimum of 25 to 30 leaves per apple is suggested on M.26, Mark or M.7A rootstocks; probably fewer are needed on M.9 rootstock, which is very efficient.

In baking tests, Fuji took a long time to cook and had a tough skin. I believe it to be primarily a fresh dessert apple. Whatever its faults, Fuji has one exceptional characteristic; i.e., long shelf life at room temperatures. Even if not kept in cold storage for a week to two weeks, it seems to develop better flavor and stay crisp.

As local market prices come down, one wonders if commercial growers can make a profit with it because of the high cullage mentioned. In this valley, it would not necessarily be unmanageable, but it does exhibit Jekyll-and-Hyde tendencies.

Gala

✳

Tree:

   Synonym: Kidd's D-8. Now has a number of striped and redder strains. **Origin** was New Zealand, by plant breeder J. H. Kidd. Crossed in 1934 and named in 1950. **Parentage** is Golden Delicious x Kidd's Orange Red. **Diploid** of 34 chromosomes. Tree has high vigor but is medium-large in **size** and tends to over-produce. **Annual** bearing if well thinned early. **Winter hardiness** is good in USDA zones 5 through 9, with trial suggested in protected lower zone 4.

   Disease susceptibility: to fireblight, apple scab fungus and somewhat susceptible to powdery mildew. In wet, mild winter climates, such as my area, it can get anthracnose fungus. **Sources** are Stark Bros., Carlton Plants, Van Well Nursery and other nurseries. Most nurseries carry the redder strains.

Fruit:

   Size is medium to sometimes small. M.9 rootstock helps give larger apples, but good orchard management is important. **Shape** is round-conic to oblong-conic, similar to a small Golden Delicious. Occasionally slightly ribbed. **Skin color** is golden-yellow with reddish blush. Royal Gala is darker with dark red, rather narrow stripes. **Stem** is thin, variable and usually well above the cavity. Some are long stemmed. Does not seem to **russet**.

Quality rating: **Very good to Best** in most apple growing areas; flat tasting in very cool summer climates.

◇ ◇ ◇ ◇ ◇ ◇ ◇ ◇ ◇ ◇ ◇ ◇ ◇ ◇ ◇ ◇

   It is necessary to go back to the grandparents in order to clear up some misconceptions about its parentage. One parent, Kidd's Orange Red came from a cross of Cox's Orange and Red Delicious around 1924, so Cox and Red Delicious are Gala's grandparents on one side, not its parents as some have stated.

   Kidd's D-8, plant patent #3637, is the standard or original Gala. The Royal Gala, Tenroy cultivar, plant patent #4121, is a redder, more heavily striped strain. I have experience in growing both, and I prefer the original. Stark Bros. obtained the U.S. patent rights some years ago, but in recent years they have given other nurseries special licenses to grow and sell the two patented strains.

Despite the recognition and assertion by Stark Bros that this was a very special apple, originally they could hardly give the trees away. Then, New Zealand came to the rescue. With no promotion, New Zealand shipped standard Gala apples to port cities on our West Coast starting in 1981. I was then able to find Gala apples in a number of stores. Originally, they sold for what seemed like high prices, from 89 to 98 cents a pound. Each year, increasing amounts of this cultivar are exported to North America, yet the prices have gone up to $1.49 a pound in many stores.

Growers who have a small acreage, grow quality fruit, and sell direct to the consumer have the best control of their economic destiny. They are doing very well with Gala here in the Willamette Valley.

With the seasons in New Zealand being the reverse of ours, their harvest starts in February and ends after about three to four color pickings. Gala begins to appear in the stores in western Oregon in March to early April. My own Gala, in western Oregon, matures in early September.

On the Columbia River near Orondo in eastern Washington, commercial orchardists pick Gala starting about August 20, when a few apples exhibit cracking around the stem. This is their guide for the first picking in that area. More than one color picking is necessary on the standard Gala. I have not seen this type of cracking around the stem in my orchard, and believe their higher temperatures and faster growth contribute to this problem.

Gala has viable pollen for other cultivars but it should not be pollinized by either of its parents. It appears to be a low chill cultivar, requiring 800 or fewer hours between 32° and 45°F when dormant.

One of the great attributes of this cultivar is its ability to stand fairly high temperatures without excessive sunburn or poor quality. It is starting to be grown in California. Gala fills the period just before the late apple season.

The tree's limbs are brittle and, with a heavy crop, they break easily in the wind. I suggest using a trellis or some type of support system.

This is a delightful apple. Some are small but most are medium in size; none are large. As prices come down and production goes up in the U.S., commercial growers may find it more difficult to make money with Gala because the poundage per tree is lower than with larger apples. Gala apples are considerably smaller than most cultivars, and apples are sold by the pound, not by size. This cultivar requires good management in order to grow it profitably. Let's hope it will not be over-produced in the U.S. and hurt the growers.

Ginger Gold™

✳

Tree:

Synonyms: called the Mt. Cove strain with Plant Patent #7063, held by Adams County Nursery. A chance seedling of unknown **parentage**. A **diploid**, 2 x 17 = 34 chromosomes. **Size** of tree is large and vigorous, with wide angle branches. **Very productive** on young trees (3 to 5 years); there are no old trees as yet. **Annual** bearer in Virginia if thinned early. **Winter hardiness** is still unknown as it has fruited primarily in USDA zone 6 in Virginia and zone 6 in Washington State, but it is hardier than Golden Delicious.

Disease susceptibility: powdery mildew fungus; somewhat susceptible to apple scab fungus. Moderately susceptible to fireblight. Blossom time is in the mid- to late-bloom category and exhibits some tip bearing (branch tip bloom.) **Sources** as of this writing are Adams County Nursery and Van Well Nursery (see "Nursery Addresses.") Others may be licensed soon.

Fruit:

Size is medium to fairly large. **Shape** is mostly conic, somewhat like Golden Delicious; some are more round-conic. **Skin color** is green before approaching ripeness, very yellow when ripe. **Stem** length is medium to quite long. **Flavor** is very good, with a surprising amount of sugar. **Texture** is crisp, firm, juicy and slightly coarse, with yellowish-white flesh. In Virginia, **russets** sometimes around the stem; no russeting has been noticed in eastern Washington. **Matures** July 25 to August 15, depending on locality. Exact **storage time** is not known, except it appears to keep at least two months at 30° to 32°F. No **physiological disorders** have been noted. My advice is to sell it as soon as it is sweet, in August, when it competes only with long-stored apples many months old.

Quality rating: **Very good.**

✧ ✧ ✧ ✧ ✧ ✧ ✧ ✧ ✧ ✧ ✧ ✧ ✧ ✧ ✧ ✧ ✧

This is one of the newest apples described and pictured in this book, and was released in 1982. Patenting an apple today is often not worth the time and money, but officials at Adams County Nursery told me that it appears there is no apple grown in the U.S. that ripens this early, tastes this good, and yet keeps for two months with quality in regular cold storage. It is a summer apple that acts like a fall or winter apple. This is very rare.

After hurricane "Camille" tore up Clyde and Ginger Harvey's orchard in Virginia, from the fractured roots of a grafted Winesap came this fine new early yellow apple. There was some speculation that the roots were seedling Newtown, open pollinated, but this has not been proven. Its origin is quite a mystery. Then the rootstock, planted from seed, became a great new cultivar. Ginger Gold™ appears to be one of those defying the odds; not many seedlings produce worthwhile fruit.

We should remember that of the top 15 commercial cultivars in the U.S. as of 1990, only two, Idared and Cortland, come from research stations. The other 13 were chance seedlings some amateur or commercial grower brought to the attention of the public or a nursery.

My first crop matured in August 1994. I am impressed that this early apple takes pre-harvest heat so well, as it does in Virginia and in the Yakima Valley of Washington. If you like such an early apple, here is another one for the 21st century. I think it is somewhat pointless to store it and have it compete later with Gala and others. It is much superior to a green and sour Golden Delicious stored eight months.

A niche apple for the early season.

Golden Delicious

✳

Tree:

Origin is discussed below. A **diploid** of 34 chromosomes. **Size** is medium. **Productivity** is very high but biennial if not thinned early. **Winter hardy** in USDA zones 5 through 8, but has been grown in parts of zones 4 and 9. **Disease susceptibility:** to scab, mildew and certain cankers. **Bloom** period is mid to late. **Sources:** many commercial nurseries.

Fruit:

Size is medium to medium-large, if thinned. **Shape** is conical. **Stem** is quite long and usually slender. **Flavor** is sweet and spritely. **Russeting** often occurs from cool drizzly weather after blossom drop or from mildew. When picked ripe, **storage** is two to three months in regular storage at 30º to 32 ºF.

Quality rating: **Very good to Best.**

❖ ❖ ❖ ❖ ❖ ❖ ❖ ❖ ❖ ❖ ❖ ❖ ❖ ❖ ❖ ❖

This cultivar has a beautiful and appropriate name. The apple is golden yellow in color and delicious in taste. If picked green for long storage, it is neither golden nor delicious. If picked after allowing enough time for the characteristic yellow coloring to appear, it is one of the best all-purpose apples in the world. It reaches ideal quality in the Willamette Valley when and if the rains start in late September or early October. During this time, it is cool enough so that Golden Delicious ripens slowly and stays on the tree (few, if any, fall off.) Under these conditions, by the time late October comes around, this apple's color is now a full yellow and it is optimum picking time. The taste will be at its absolute best even if the apple is picked somewhat late for storage.

The waxy finish decreases shriveling in storage. Without this finish apples do tend to shrivel, and high humidity is needed in the storage area. When dealing with a small quantity, one can store the apples in the home refrigerator in plastic bags with pinholes in the plastic. This reduces the chances the apples will dry out, and storage life is increased.

If the Golden Delicious is grown in areas that are fairly warm and have dry summers, as in the northwestern U.S., especially eastern Washington, the fruit will have a waxy finish and a beautiful color and appearance. This cultivar is also grown in parts of Italy, Spain, and most other apple producing areas, including the southern hemisphere.

Golden Delicious has superior tree characteristics, and adapts to many climates and soils. The characteristic of adaptability has been very important over the last 200 years in determining the commercial popularity of Golden Delicious, Red Delicious, Ben Davis, Jonathan and others. Golden Delicious is now grown worldwide, and it is in second place in production of all cultivars grown.

The Golden Delicious is not as popular as it once was, because packers and buyers have been trying to make it a seven to eight month storage apple by demanding that it be picked while still green, with very little or no yellow color. This is done in spite of the well-known fact that this cultivar's soluble solids (sugars) do *not* increase, or increase only slightly, if the apple is picked when light green in color. In contrast, Red Delicious and Newtown show an increase in soluble solids when stored. In other words, a Golden Delicious picked green stays a green, sour Golden Delicious. Excellent taste can be, and should be, a feature of Golden Delicious, provided it is not picked too early.

In the early 1900s Anderson Mullins owned a 36-acre farm in the hill country of Clay County, West Virginia. His father had planted some apple trees in the 1880s named Golden Reinette. It was a very old English cultivar, and the fruit size was small. Golden Reinette ripened in October and, where it was exposed to the sun, often exhibited russet spots and some red strips on a golden yellow color. Its flavor was highly esteemed in England, and it was probably in the **Very good** category.

Another apple, the Grimes Golden, was also planted in the area, and many people think that the first Golden Delicious tree grew from a seed of Grimes Golden that had been pollinized by Golden Reinette. Many seedlings sprang up in this primitive orchard, but only one was outstanding. Apparently this new find fruited for nine years before Mr. Mullins realized he had something special that the world should know about. In 1914 he mailed three Golden Delicious apples to Paul Stark of Stark Bros. Only a few years before this, the Starks began selling the pointed and distinct Delicious cultivar, the sports of which are now called Red Delicious. Since this new golden beauty was long, conical and pointed like Red Delicious, it was named Golden Delicious, but it has no close relationship to the Red Delicious.

Stark visited Mr. Mullins' farm and found that the Golden Delicious tree was hardy, healthy and gave heavy crops. He immediately offered Anderson Mullins $5,000 for the tree and 30 square feet of ground around it. If Mullins accepted, he was to care for the tree, pruning, spraying and fertilizing it with manure. Since $5,000 was quite a sum for the time, and since the Stark Bros. Nursery was famous, Anderson Mullins accepted the offer. Eventually, the Stark Brothers would sell millions of Golden Delicious trees.

The tree probably first sprouted some time between 1897 and 1901. It first fruited in 1905, and then lived until 1958, almost 60 years.

Grimes Golden, probably one of the parents of Golden Delicious, usually beats Golden Delicious in taste tests; however, its tree characteristics are inferior to those of Golden Delicious. The Golden Delicious cultivar gives acceptable fruit from the southeast corner of Minnesota down to the Mason-Dixon line and below. Even the Los Angeles Basin and northern Florida have fruited it, but I do not know with what quality. It is productive, healthy, and grows in many soil types, has viable pollen and has a long bloom period, from mid-season to late-season. It is, therefore, one of the best varieties to use to pollinize other cultivars that bloom at the same time, with one exception—its first generation offspring. Do not use Golden Delicious to pollinize first generation offspring, as we know its pollen will not pollinize Jonagold and probably not other offspring. In comparison, Grimes Golden only does well in certain climates and soils, it is often biennial in bearing fruit, and it frequently gives small apples.

Golden Delicious does best when irrigated and grown in areas where the summers are warm and dry. In my orchard it does not ripen well until mid-October, often later, if the cool rains have started. Many people have never tasted a tree-ripened Golden Delicious apple, but it is superior when grown correctly.

Most of the spur strains of Golden Delicious first appeared in 1955 in eastern Washington after a very severe and damaging early November freeze, when the trees were not yet fully dormant. This certainly suggests that early very cold weather can either cause or facilitate bud mutations. Since Golden Delicious oversets unless carefully thinned early, the spur strains were not very valuable, except that these trees were 20 to 25 percent smaller in size, which some growers preferred. The fruit of all the spur strains appears inferior to that of the standard Golden Delicious cultivar; therefore, I do not recommend them.

Granny Smith

✳

Tree:

Origin is as an open pollinated seedling of French Crab, a large green apple it much resembles. Found in Marie Smith's orchard near Sidney, Australia. Chromosomes are 2 x 17 = 34, a **diploid**.

A vigorous and medium- to large-**sized** tree. Bears early and heavily, on tips of branches as well as along the branch. May have some blind wood (no fruit spurs). **Annual** producer that goes into dormancy late. **Winter hardiness** is somewhat unknown, but I suggest planting in USDA zones 6-8, with trial in some of zone 9. **Disease susceptibility:** to apple scab fungus; very susceptible to powdery mildew fungus. **Blossom** season is in the mid- to late-bloom period.

Sources: most commercial nurseries. I suggest using only the original strain.

Fruit:

Size is medium to medium-large. **Shape** is round-conic to round. Color is usually a bright green with an occasional brownish-red flush. **Stem** length is medium to long, and usually slender. **Flavor** is very tart when first picked, but it does gain sweetness after storage. **Flesh** is white, sometimes tinged green. **Texture** is very firm and juicy with a thick, tough skin. It seldom if ever **russets**. **Matures** in late October in eastern Washington, but in early November here in the Willamette Valley of northwestern Oregon. **Storage life** is four to six months in ordinary cold storage, longer in C.A. **Physiological disorders** that can occur are bitterpit and skin scald in storage. May water-core if picked too late.

Quality rating: **Good to Very good.**

◇ ◇ ◇ ◇ ◇ ◇ ◇ ◇ ◇ ◇ ◇ ◇ ◇ ◇ ◇ ◇

Despite the fact that Granny Smith has been sold in North America for some years, many consumers think of this apple as new. While new to some consumers, it was discovered about 125 years ago. An accurate history of this cultivar is recorded in Washington State Extension Bulletin #0814, from which I have made this excerpt. It was written by (then) extension agent Jim Ballard and published by Washington State University in 1981.

The woman responsible for this apple never received much appreciation for giving the world the fruit that bears her name. Marie Ann (Granny) Smith was the wife of Thomas Smith. The Smiths emigrated from Sussex, England, in 1839 and developed a 5-acre orchard in the Sydney suburb of Ryde, Australia. One morning in the year 1868, Mrs. Smith asked E. H. Small, a successful fruit grower, to look at a seedling apple that was growing in the orchard and to express an opinion about it. His son, Tom Small, then a 12-year-old, went along with them. The tree was growing among ferns and bladey grass and had a few very fine specimens of green apples on it. Mr. Small tasted it critically, and remarked that he assumed it would be a good cooking apple, and might be worth grafting from, though Mobb's Royal and several other good cookers seemed to fill the demand at that time. Tom, however, remarked that it was also a good eating apple as he sampled it.

Mr. Small asked "Granny" how the apple came there. She replied that she had brought gin cases back from the Sydney markets which had contained some Tasmanian apples which were rotting. She said she had thrown them out at the site where the original Granny grew. It is thought that Mrs. Smith mentioned that the remains in the cases were of French Crabs. The greasy skin and keeping qualities of "Granny Smith" point to this being correct. Mrs. Smith grafted a few of her trees, and not long afterward Edward Gallard, another member of the family, planted out a fairly large block of these "Grannys" and is credited as being the first commercial grower of the new variety.

Early tests in the Bathurst district west of Sydney were completed shortly after the turn of the century. After its excellent storage ability was established, the fame of the variety spread. Granny had met the full test:

- It was a grower's apple.
- It satisfied the packer and shipper.
- It stood up well in the markets.
- And finally, the consumer liked it and asked for more.

The first publicized spur-type mutation of Granny Smith was found in Australia. This Hannaford strain found near Adelaide has been widely researched in many countries and was found, in general, undesirable because of virus content, tree growth characteristics, and fruit quality not up to the fruit from virus-free standard Grannys.

Late ripening is a characteristic of this apple, and it requires a long warm fall to ripen. The areas of New Zealand, Australia, South Africa, Chile, southern France, southern California, eastern Washington and the Willamette Valley of Oregon are among the best growing sites.

Eastern Washington has some areas in USDA zone 5 and, more years than not in these areas, Granny Smith apples freeze slightly on the tree before being picked. These light freezes usually do not prohibit the apple from being marketable, but obviously that area is its very northern limit for growing in the West.

Granny Smith apparently has a low chill requirement, as friends of mine have seen apples growing in a New Zealand garden alongside a pineapple plant. Much of southern California is in a low chill area, and they produce a rather large acreage of Granny Smiths; however, in many years it sunburns and colors in the hot San Joaquin Valley unless given considerable shading by limbs and leaves.

Before apple researcher Dr. Donald McKenzie of New Zealand was tragically killed in an auto accident, he graciously accepted an invitation to address us in a field demonstration sponsored by the Willamette Valley Fruit Tree Growers Association. I clearly remember him discussing Granny Smith; he said a mature orchard planted with the original strain had produced 4,000 boxes on an acre, and that production of 2,000 boxes an acre was not uncommon. He also said the first spur strains from the virus-infected Hannaford strain were inferior and were not planted any more. In other words, his advice was to stay with the original Granny Smith.

In storage, it can develop a greasy, tough skin, so some people prefer to peel it before eating it. Its fine shipping, long storage and home shelf-life as a firm and juicy apple no doubt save this one, as I don't think it has the flavor or sugar of some of our choice apples.

Growers all over the world are paying off mortgages with this old-timer, regardless of its drawbacks; however, apple prices have begun dropping for most commercial growers, so new plantings will probably soon cease.

Gravenstein

✳

Tree:

Parentage is unknown, and it has three sets of chromosomes, 3 x 17 = 51. A **triploid** that is not suitable for pollinizing other cultivars. **Vigor** is high and it becomes a very large tree even on semi-*dwarfing* M.26 rootstock. It is *not* an early bearer unless grafted to the smallest rootstocks (often one to three years later than many cultivars.) **Biennial bearing** is one of its traits disliked by many growers. Grown commercially in very few areas except western parts of Washington, Oregon, and California, usually west of the Cascades.

Disease susceptibility: to most everything including apple scab fungus, canker, powdery mildew and fireblight. **Blossom** season is quite early in most zones, except the Rosebrook strain. Other early bloomers that pollinize most strains well are Erwin Baur and Idared. Early blooming crabs are also effective. **Sources:** widely available, but some strains may be seedlings of poor quality. Plant a known strain.

Fruit:

Size is medium to large. **Shape** is quite variable and usually ribbed. Others are often round and even elliptical. **Skin color** is very dependent on picking time and will vary from light green to striped red over yellow (see picture.) Stem is usually short but some go to the top of the stem cavity. It **russets** sometimes around the stem. **Flesh** is white to ivory in color; **texture** is tender, crisp and juicy with a most appealing tart **flavor**, both for cooking and fresh eating. **Matures** between August 20 and September 15 in most climates. **Physiological disorders** are sometimes bitterpit and water core; calcium deficiency is thought to cause both.

Quality rating: **Very good to Best.**

❖ ❖ ❖ ❖ ❖ ❖ ❖ ❖ ❖ ❖ ❖ ❖ ❖ ❖ ❖ ❖

This is a very old apple variety of unknown parentage. We first find a definite record of it in old Germany (now Denmark) around 1670.

Gravenstein may have originated in the garden of the Duke of Augustinberg in Schleswig-Holstein. Other accounts suggest it to be the very old Ville Blanc from Italy, with scions of this cultivar having been brought to Grasten Castle, South Jutland, by a brother of Count Ahlefeldt (Beach, 1903.) Others suggest that Gravenstein is from northern Russia, but that seems doubtful as it does not seem hardy enough. Since the tree is not very hardy and is subject to sunscald on cold, sunny winter days, I lean toward believing in the Italian origin for Gravenstein. Apple addicts may discuss these historical anomalies over a glass of apple cider and arrive at only two firm conclusions: Gravenstein is quite old and it surely has a rich, tangy flavor. The cultivar was brought to eastern North America from Europe about 1826, and northern Californians told me the Russians brought it there before 1820.

Because of such unusually high quality, it was rated "very good to best" by Beach, a high rating given to only a few apples of the 700 or so included in his book, *Apples of New York*. For applesauce and pie, Gravenstein has rabid followers who would almost mortgage the farm for a couple of boxes of the apple about August 30. Many farmers still grow their own trees where climate permits. Even in Portland, Oregon, many city lots have a huge old tree, some nearly a century old, producing bushels of the fruit. When striped and nearly ripe, the variety's fresh dessert flavor is rivaled only by Elstar that early in the season. Gravenstein's quality is all that saves it from extinction, as it is a most frustrating apple to grow commercially and make a profit. The trees usually contribute only marginally to an orchard's profitability.

Classified as a biennial bearer, an understanding of its genetics can lead to getting some apples almost every year. In the first place, Gravenstein is a triploid with 17 extra chromosomes (51 total), which means it is nearly sterile, incapable of very much self-pollination, and does not produce enough viable pollen for other cultivars. In pioneer times, it was typically planted with the old Tompkins County King, another triploid of fine flavor. This is probably why many pioneer orchards, which were often isolated from the viable pollen of other varieties, produced such stingy amounts of fruit and gave both varieties the bad reputation of being shy or intermittent bearers. Remnants of these old orchards still exist in the Willamette Valley.

The East Malling Research Station near Kent, England, suggests that triploid varieties should have pollen from two viable pollinizers to properly set seeded fruit and to pollinize each other as well as the triploid.

Yellow Transparent also catches the early bloom, but I think it is so inferior to Gravenstein that I have placed it in a special quality class called "one-half hour apples," (meaning use them one-half hour after picking.) In a commercial orchard, Idared and Liberty are both good saleable apples, so I would suggest one of these pollinizers or, in a home orchard, use Erwin Baur. Jersey-Mac has also been used as an early pollinizer, but frequently it does not bloom early enough and, in my opinion, is an inferior apple.

Gravenstein has a fairly long bloom period, depending on how cool and drizzly the spring weather is. Its later blossoms may open during mid-bloom, when any number of varieties will help pollinize it, but that does not always occur. Because in our weather about two to three weeks go by from first bloom until the last, picking time follows a similar pattern. The first picking falls in the period of August 15 to 30, while the last picking may be as late as September 15 or 20.

The reader will note from the color illustration of the Gravenstein cultivar that a given tree will produce the striped, ripe apple from the early bloom and produce the green apple from the late bloom. The long picking time also makes Gravenstein more expensive to grow commercially. It has a tendency to drop much of its fruit and, unless thinned properly, the fruit tends to grow in clumps of two or three. A few years ago, I foolishly volunteered to help spray and pick apples in an 80-year-old orchard containing mostly Gravensteins and Kings. The ladder was inadequate for picking so high from the ground, and the apples had not been thinned. I managed to pick one apple at a time while hanging onto the ladder with the other hand. Also, due to its short stem, the other apple (or other two apples) would move downward, break off, and fall 15 to 20 feet with a bruising thump which left them fit only for cider. One is fortunate to last through such an experience without serious injury, and a complete overhaul of one's life and accident insurance is in order. As a result of this experience, I devised a different pruning plan for these lovely old giants. Why not prune them with a chain saw at ground level and make some beautiful furniture? I subsequently made this half-joking proposal in a lecture with over 100 people present and now have a few new enemies.

Only some of the other triploid cultivars such as Baldwin and Mutsu, rival the Gravenstein tree in size and vigor. I suggest that Gravenstein not be grafted to any rootstock larger than M.9, M.26 or M.7A unless one is willing to work 20 feet above ground in a few years.

With cold and rainy spring weather, most any variety can form fruit without fertilization. Lack of fertilization is probably induced by cold weather at bloom time, but the actual agents preventing fertilization are plant hormones. Fruit set without fertilization is a complex physiological process which is not completely understood. Gravenstein seems especially adept at forming fruit vegetatively with no seeds, or only one or two. Hormone flow and seeding have an effect on biennial bearing. My opinion is that these *parthenocarpy* seedless fruits are not as flavorful as seeded fruits, nor do they keep as well or have as much calcium content; hence bitterpit sets in.

There also seem to be many more strains of Gravenstein than in most old apples. If, after eating a fine flavored Gravenstein, you decide to buy a Gravenstein tree, don't buy just any tree. Locate the orchard where the apple came from; or, better yet, the exact tree. Then propagate from it.

I know of at least four above average Gravenstein strains (as well as some poor strains) in the west coastal states of California, Oregon and Washington. One of the strains seems to bloom later than the others; the Rosebrook is reputed to be a mid-blooming strain, easier to pollinize. I had a Mead Red strain that seldom set much fruit until I introduced some early pollinizing trees in my orchard. Nova Scotia, Canada, produces considerable quantities of Gravenstein and, according to Beach, has a red strain discovered in the 1880s.

Gravenstein does not give its best fruit where there are consistent late July and August daytime temperatures in the 90° to 105°F range, such as in eastern Washington. For those living in the northern tier of the United States where the Gravenstein is not regarded as hardy enough, I would suggest experimenting by planting a Haralson on a cold-hardy rootstock such as Columbia, Dolgo Crab or the Polish series. Then, in about the fourth year, top graft the Haralson with Gravenstein. The Haralson just may impart enough cold hardiness from the trunk and lower limbs so the Gravenstein grafts will live through the winter. Painting the trunk and lower limbs with white latex paint would help to prevent sunscald. This is an idea that, perhaps, our researchers would not waste time on, but we don't have the same restrictions they do. Of course, it might not work!

If the apple has exceptional quality and multiple usage, it probably will survive for centuries by grafting. One seldom sees a full orchard of the Gravenstein cultivar, except in the three west coast states, and only in the western parts of those.

For 400 years or more, people from many countries have loved the flavor of the old Gravenstein, and it may be with us for many more hundreds of years.

Grimes Golden

Tree:

An old apple, its **origin** is West Virginia a few years before 1800. **Parentage** is unknown. It has 2 x 17 chromosomes, a **diploid**. Tree is medium to small in **size** on the dwarfing rootstock. It tends to be an **irregular bearer**, and some years has poor production. It is seldom a huge bearer of good-sized fruit. **Winter hardiness** is poor and apples do not size well north of lower USDA zone 5. I suggest growing in USDA zones 6 through 8.

It is susceptible to the **disease** collar-rot if the Grimes trunk is at or below ground level. The Grimes graft union should be at least two inches above ground level and, since M.9 root is quite resistant to collar rot, that is the rootstock to which mine is grafted. Slightly susceptible to apple scab fungus. **Blossom** season is from the early- to mid-bloom. Some years it is early enough to pollinize Gravenstein.

Sources are many nurseries in the eastern U.S. and some in the west. Southmeadow Fruit Gardens in the midwest also propagates it.

Fruit:

Size is usually medium to medium-large in favorable climates. **Shape** is round to round-conical. On the **calyx end** there are usually five distinct crowns in a flat, wide plane, with the basin quite broad and deep. **Skin color** is an attractive clear, deep yellow. Fruit tends to **drop** before or just at ripening. **Flavor** is exceptionally pleasing with a spicy, somewhat coarse-grained light yellow flesh. Usually wins taste tests over its offspring Golden Delicious. **Matures** in mid to late October in USDA zones 7 and 8 of the Willamette Valley, Oregon. Except for dropping, no other **physiological problems** are noted.

Quality rating: **Very good to Best.**

◇ ◇ ◇ ◇ ◇ ◇ ◇ ◇ ◇ ◇ ◇ ◇ ◇ ◇ ◇

According to Beach, the original tree was said to be fruiting in 1804 in West Virginia, so its origin dates at least to the late 1700s.

Under most conditions, commercial orchardists have described Grimes as "unmanageable" for a commercial orchard, but it was a universal favorite where it could be grown. Some of the better areas in which to grow it are the Virginias, North Carolina, western Oregon, Pennsylvania, Indiana, Illinois and Missouri.

All the Grimes Golden apples I have tasted from New York were fine, but they were small. Those from warmer Virginia were much larger. Also, in my home valley in Oregon Grimes does well; however, it is a variable bearer and one does not know from year to year if there is to be much of a crop. It might give three crops in a row and then not much the fourth year. It appears to do best with a pH of 6.5 to 7.0.

I sent my first box of Grimes (from M.9 rootstock) to two of our children in college in Minnesota a few years ago, and they raved about the taste. Now I am down to one tree to keep reminding the family that Grimes is a prize with high flavor and great taste, even better than its offspring Golden Delicious.

Only 20 miles north of me, across the river from Portland, Oregon, in Vancouver, Washington, there is a small direct-market apple grower who has a considerable number of Grimes Golden trees. He says half of his crop is sold out yearly before the apples are ripe.

If you want a little more richness and tang than Golden Delicious, and you have a warm climate, perhaps you should try growing Grimes Golden. It could become a family favorite.

Haralred

✳

Tree:

Synonym: plant patent #4824. A limb sport of Haralson with 2 x 17 = 34 chromosomes, a **diploid**. **Size** of tree is medium to small, but with heavy **productivity**. An **annual bearer** if thinned. **Winter hardy** into parts of zone 3; is outstanding here in USDA zones 7 and 8 in Oregon. It has **disease resistance** to fireblight. **Bloom** period is from late-early-bloom to mid-bloom.

Source of the tree is the patent holder, Bailey Nursery.

Fruit:

Size is medium to sometimes small, variable depending on culture management. **Shape** is usually conic to round-conic. **Skin color** is pink-red with round gray-brown conspicuous dots. **Stem** is usually medium in length. **Flavor** is an appealing sweet-tart. **Flesh** is white; **texture** is crisp, firm and fine-grained. **Stores** well for three to four months. Seems nitrogen sensitive with excess application.

Quality rating: **Good to Very good.**

◇ ◇ ◇ ◇ ◇ ◇ ◇ ◇ ◇ ◇ ◇ ◇ ◇ ◇ ◇ ◇

Haralson is an open pollinated seedling of Malinda. One can only speculate what the other parent was. Malinda was found in Vermont, of unknown parentage. It was sent to Minnesota about the time Peter Gideon planted the seeds of cherry crab at Excelsior, Minnesota, and propagated the Wealthy cultivar in 1860. Around the turn of the century, the University of Minnesota planted a great many seeds of Malinda because it is very cold hardy. Beach, in *The Apples of New York*, does not give Malinda high marks, but he also added that the cultivar had not yet been tested in New York, an anomaly one occasionally finds in his two volumes. He said it was very cold hardy.

The Haralson cultivar was selected in 1913, tested as Minnesota 90, and was rather quickly introduced in 1923. Since our Minnesota farm was only 30 miles south of the Excelsior Research Station, I remember eating Haralson as a boy. This new cultivar was quickly established on most of the farms in that area.

This north country, very hardy apple is somewhat maligned by those who live in a milder climate and can grow the late varieties that ripen in climates like mine. In the northern one-third of Minnesota (zone 3), Haralred is picked in late September, about one week earlier than Haralson, before the onslaught of the hard freezes that frequently come from October 20 to November 15.

The University of Minnesota lists Haralson's uses as fresh eating, pie, sauce, baking and freezing. It keeps well in cold storage until March if not picked too late. Haralred was found as a limb on a Haralson tree in southeastern Minnesota and is probably a true mutation.

Of the 99 commercial Minnesota orchards surveyed in 1983, there were almost 75,000 Haralson trees in the ground on all types of rootstock. This is more than double the number of the second place Fireside and the Connel Red strain of Fireside. Of the over 300,000 commercial apple trees, about 25 percent are Haralsons.

Haralred matures a few days earlier than the original Haralson, and its flavor is somewhat sweeter. Haralred does not seem to russet as much as Haralson sometimes does, and it appears to have the same fireblight resistance. Bailey Nurseries, Inc. was granted plant patent #4824 for Haralred.

This is a true north-country fruit and it will succeed where many others will not if grafted on cold-hardy rootstock. The University of Minnesota recommends it for all four special zones of Minnesota, and it is a productive and annual bearer.

This is the only apple cultivar where I prefer the red strain (Haralred) over the original. As grown here in my area on M.9 rootstock, Haralred is an exceptionally attractive dark pinkish-red, with small yellow dots on the skin. It does well in taste tests against other fine tasting apples.

Hudson's Golden Gem

✳

Tree:

   Origin: as a seedling in a fence row thicket at Tangent, Oregon, about 1925 to 1930. Parentage may be seed from Golden Delicious, open pollinated. It is a diploid of 34 chromosomes. Tree size is medium, but quite vigorous. Annual bearer if thinned. Winter hardiness is unknown, but I suggest USDA zones 5 through 9.

   Quite disease resistant to apple scab fungus, powdery mildew fungus and fireblight. Sources: a number of small specialty nurseries such as Rocky Meadow Nursery, Southmeadow Fruit Gardens, etc.

Fruit:

   Size is medium to medium-large. Shape is long and conic. Skin is an attractive light grey-gold russet over all the apple, very different. Stem is usually quite long, similar to Golden Delicious. Flavor is very good, with a sweet, tart, pear-like flavor. Matures here in our USDA zones 7 to 8 between October 10 and 30. Stores for two to three months at 32°F.

Quality rating: Very good.

◇ ◇ ◇ ◇ ◇ ◇ ◇ ◇ ◇ ◇ ◇ ◇ ◇ ◇ ◇

   Hudson's Golden Gem was introduced by A. D. Hudson in 1931 through Hudson's Wholesale Nurseries of Tangent, Oregon. Tangent is a village a few miles from Oregon State University at Corvallis, Oregon, in the middle-southern part of the Willamette Valley.

   This apple had only local dissemination until after World War II. Now it is surprising how many connoisseurs have a tree in their collection. Promotion by Robert Nitschke of Southmeadow Fruit Gardens no doubt had much to do with it. Promotion, of course, does little good unless the apple is above average, and this one is. One reason for growing it is its resistance to apple scab fungus, at least to our strain of the fungus.

This book only features three full russets, Roxbury, Ashmead and Hudson's Golden Gem. There are at least 15 very fine-tasting russets in the world, mostly from Europe, but many are difficult to buy in this country. Most good-tasting russet apples are old, and a nostalgic holdover from an earlier time. Hudson's Golden Gem is probably the newest of this group, being only 60 years since found.

At a fruit show some years ago three friends asked for some apples. I put four different cultivars in each of three paper bags and, in addition, I slipped three Hudson's Golden Gems in each bag without letting them know. There was an immediate phone response; all wanted to know what that wonderful-tasting, strange-looking apple was.

Its big fault is apple cracking in juvenile years, as with Cox Orange. Spraying for three or four weeks from blossom time (two sprays) with the plant hormone GA-4-7 helps, plus a followup of three to four calcium chloride sprays (once a month).

Only a few miles from my home, there is a Hudson's Golden Gem tree which is about 40 years old. My new tree is from a graft of this old tree. One-half of my apples cracked by June 15th the first two years, but none crack now. I have never seen cracked apples on the old tree, so I assume it occurs only in the younger trees.

I get considerable mileage as a conversation piece from this unique fruit. It is a high quality apple, and it also puts some of Oregon history in my yard.

Idared

*

Tree:

   Origin: crossed in the 1930s at the Idaho Agricultural Experiment Station from Jonathan x Wagener. Selected in 1935. It has 2 x 17 chromosomes, a **diploid.** Small to moderate tree **size,** but productive. I suggest USDA zones 4 through 6, except it also does well in the Willamette Valley zones 7 and 8. **Disease:** susceptible to fireblight, scab and mildew. An early **bloomer,** before most commercial apples.

Fruit:

   Size is medium-large, sometimes quite large, unlike its parents. **Shape** is round to round-conic. **Skin color** is a bright, light red over green. **Flesh** is white, sometimes tinged green, and fine and crisp. Usually **russet-free. Matures** between October 5 and 25 in most USDA zones. **Stem** is rather short and slender, usually to the cavity top. **Physiological disorders:** seems to have none.

Quality rating: **Good to Very good.**

❖ ❖ ❖ ❖ ❖ ❖ ❖ ❖ ❖ ❖ ❖ ❖ ❖ ❖ ❖

   Idared is essentially a commercial cultivar, and the newest to make the top 10 in the U.S. Leif Verner of the Idaho Experiment station in Moscow, Idaho, was responsible for this cross of two excellent old American apples.

   Since the 1940s, it has slowly climbed in popularity with commercial orchardists and in 1993 5 million boxes were produced in the U.S. It is becoming more popular in Northern Europe.

   In Oregon, Idared blooms as early as Gravenstein, and it is the primary pollinizer for Gravenstein. Very few apples that ripen late bloom this early. Apples stay on the trees well until they are thoroughly colored.

   The Idared tree is of small to moderate size, which is not surprising since one of its parents, Wagener, is quite genetically dwarfing, and its other parent, Jonathan, is of only moderate size.

This apple generally tastes better after being stored for a month, and it is one of the best keepers in cold storage, remaining good until March or April. Home orchardists find that its attractive color, annual bearing and good keeping offsets a rather tart quality. A multi-purpose apple used for sauce, pie and fresh dessert.

For a grower who wants good quality and ease of growing, Idared is a fine choice; however, this cultivar, like Jonathan, is quite susceptible to fireblight—which we are not concerned about here in our valley. Other areas, of course, are.

Jonagold

✷

Tree:

   Parentage: a cross of Golden Delicious x Jonathan. A **triploid** of 51 chromosomes, so do not use as a pollinizer. It is *not* cross-fruitful with Golden Delicious, one of its parents. Once it has started to fruit on size-controlling rootstock, it has only moderate **vigor** and **size**. **Winter hardiness** is unknown, but USDA zone 5 is probably its northern limit. I suggest growing it in zones 5 through 8.

Fruit:

   Size is large to sometimes very large. **Shape** is usually conic, like a large Golden Delicious. **Skin color** is a yellow background with a pinkish-red blush and/or stripes. **Stem** is usually long, similar to Golden Delicious. **Flavor** is sweet-tart, but it is regarded as a sweet apple by most. It is not prone to **russet**, but it can in cool, drizzly spring weather with mildew present. **Matures** October 5 to 15 in western Washington and western Oregon. **Storage life** is medium to short. **Physiological disorders:** susceptible to bitterpit. Needs four to seven calcium chloride sprays to help prevent the calcium deficiency of bitterpit. Will sunburn in hot climates. Susceptible to spring frosts at blossom time. Nitrogen levels should not be high; keep at leaf analysis levels of 1.7 to 1.9 when taken the first week of August.

Quality rating: **Very good to Best.**

❖ ❖ ❖ ❖ ❖ ❖ ❖ ❖ ❖ ❖ ❖ ❖ ❖ ❖ ❖

   This is one result of the long-standing apple breeding program by the experiment station in Geneva, New York. The apple was named by Dr. Roger Way. It was tested as N.Y. no. 43013-1 before being named and introduced in 1968 as Jonagold. The northern Europeans, especially in Belgium, tested it very quickly after its release. It has been one of the two most planted apple cultivars in northern Europe, at least until recently.

   Some of the first plantings in western Washington and Oregon were probably made due to the influence of Dr. Robert Norton, former superintendent of the Northwest Washington Research Station, which is located 60 miles north of Seattle at Mt. Vernon. Jonagold grows to perfection in the Skagit Valley of northwest Washington, and perhaps as well in the Willamette Valley of Oregon. It appears to be an apple for areas west of the Cascade Mountains, as it does not accept the heat of eastern Washington except at over 1,500 of feet altitude or north and northeast of Wenatchee, Washington, on into Canada.

In the blind taste tests at the 1986 "All About Fruit" Show in Portland, Oregon, the overall winner was Criterion, but Jonagold and Gala tied for second and were very close to Criterion in total count. All three of these cultivars are sweet or semi-sweet. Only the superb flavor and sharpness of the old Spitzenberg gave a sound challenge to these sweeter new cultivars.

Some of the original descriptions of Jonagold called it a long storage apple, keeping up to 6 months. My tests in storing Jonagold at 30° to 35°F show it to be an average keeper, especially if it is picked only a few days late. If I were a direct marketer, I would make sure Jonagold was sold by November 1 to 15 unless there was C.A. storage available.

Climate and management play a significant role in how this apple colors and looks. Some of the original color strains in Europe were probably more climatic strains, rather than a bud mutation.

I reject the idea that every great new apple should be solid red. Some of our most beautiful apples of the last 150 years have a yellow background with pretty red or pink stripes, providing another way for customers to identify cultivars. People who taste a prime Jonagold for the first time always become excited, and 5 minutes later do not care about its exact color. Frankly, I like the color of the original Jonagold, and with good culture in a cool summer climate, it is a beautiful apple (see picture.)

With reasonable thinning, the apples tend to be rather large—70 to 85 per 42-pound box. With excess nitrogen and overthinning, the apples become too large (¾ pound to 1 pound each), and are then very susceptible to bitterpit. Bitterpit mars the apple's appearance and shortens storage time dramatically and, in some cases, will make the apple unsalable.

Unlike some triploids such as Mutsu and Baldwin, once it starts fruiting Jonagold does not become a large tree on size-controlling rootstock. I do not need a crystal ball to know Jonagold will be grown for a long time.

Jonathan

❋

Tree:

Synonym: before 1830 it was called New Spitzenberg. **Origin:** seed parent was Spitzenberg; pollen parent is unknown. Chromosomes are 2 x 17, a **diploid.** Vigor is medium, **size** is medium to small. Leaves are narrow and rather small to medium. Branches tend to be thin and droopy. **Annual** and productive bearer if on rich soil and thinned early. **Winter hardy** between USDA zones 5 and 8.

Diseases are primarily powdery mildew and bacterial fireblight. Has some resistance to apple scab fungus. **Blossom** season is mid-bloom.

Sources are most commercial nurseries.

Fruit:

Size is medium with good management and proper thinning; otherwise, the apples could be small. **Shape** is round to slightly conic. **Color** is red with yellow-green ground color. Often a solid red, especially with the red sports. **Stem** length is medium-long, slender and above the cavity. **Flavor** is tart, aromatic and has multi-purpose use. **Flesh** is white to sometimes yellow-white and streaked pink. **Texture** is fine, tender and crisp. **Russets** around the stem on a majority of apples. **Matures** early October in western Oregon, earlier in warmer areas.

Stores well in C.A. for months. If kept in regular storage, temperature should be at least 36°F or more to help prevent Jonathan spot.

Quality rating: **Very good to Best.**

◇ ◇ ◇ ◇ ◇ ◇ ◇ ◇ ◇ ◇ ◇ ◇ ◇ ◇ ◇ ◇

This is one of the best of the important old commercial apples grown in North America. It is a fine all-purpose apple that succeeds in a wider range of climates and soils than its famous old parent Esopus Spitzenberg, although it is not quite its equal in flavor or keeping.

This cultivar is usually a moderate grower, although sometimes weak. The slender branches droop like Braeburn. It thrives best on good clay loam soil, rather than sandy soil.

Perhaps the first published account of Jonathan was in 1826 by Judge J. Buel of Albany, New York, indicating the original tree probably dated into the late 1700s. As reported in Beach, that older author (Downing) wrote that the original tree was still alive in 1845. Since most seedling apple trees live much longer than 30 to 40 years, the dating of the original tree to the late 1700s is a supportable deduction. Its seedling parent, Esopus Spitzenberg, must have been fairly well known by 1775, so the seeds which produced the Jonathan were available. Jonathan originated on the farm of Philip Rick in Woodstock, Ulster County, New York, but its date of origin was apparently unknown. Jonathan Hasbrouck first called the apple to the attention of Judge Buel, so the judge named this seedling after him.

Jonathan is more productive, more widely adaptive and healthier than its seedling parent Spitzenberg. It does well across most of North America where apples are grown.

Thanks to the advent of controlled atmosphere (C.A.) storage, Jonathan has made a slight comeback since the 1960s. Michigan leads in production of Jonathan in the U.S., followed by Pennsylvania, Washington and Ohio.

Some of the most beautiful and best tasting Jonathans I have ever eaten came from southeastern Nebraska, an area generally regarded as corn, wheat and soybean country. Midwesterners love the apple and often have to import it from other areas.

The new dark red sports exhibit a beautiful red or carmine color, often being reddish purple on the side receiving the sunlight. Wherever a twig or leaf shades the apple, a contrasting green or yellow color results, giving the apple a striking rainbow color.

The first red sport of note was the Anderson strain by Greening Nursery Company (Michigan) in 1927. I grew Anderson for many years and would have it today if a tractor had not skidded down a slope and smashed into it. Anderson may still be available from certain Michigan nurseries, and it is still my favorite Jonathan.

Jonathan is not only a fine apple in its own right, but it is one of the five or six cultivars used most frequently for hybridizing. At least 14 worthy cultivars have resulted from Jonathan crosses with another parent. The five most planted hybrids of Jonathan, arranged roughly in order of quantity planted, are:

1. Jonagold, Golden Delicious x Jonathan. Most plantings are in northern Europe, western Washington and Oregon.

2. Idared, Jonathan x Wagener. This is now in the top 15 commercial cultivars in the U.S. Grown widely.

3. Melrose, Jonathan x Red Delicious. Grown in the U.S. and northern Europe.

4. Akane, Jonathan x Worcester Permain. Some new plantings in western U.S.

5. Jonalicious, Jonathan x Red Delicious. Scattered U.S. planting. Mostly domestic (back yard.)

There is no perfect apple, which is why growers in the western U.S. should grow more cultivars than they do now. No single cultivar takes care of every buyer, every season, every color desired, every type of storage, every taste, etc., but Jonathan is one of the best all-purpose apples, especially if one likes a tart taste.

Kandil Sinap

*

Tree:

A strange and completely pyriform tree that tries to grow straight up; if not trained, it grows like a Lombardy poplar. Small in **size** even on M.7A rootstock. **Biennial** bearing unless well thinned early. Leaves tend to be small and fruit needs at least 30 leaves per fruit to help prevent pre-harvest drop. **Blooms** in mid-bloom. I suggest zones 6-9.

Source: Southmeadow Fruit Gardens.

Fruit:

Shape is long and cylindrical. **Skin color** is yellow-white with a reddish blush on the sun-side. **Flesh** is very crisp and nearly white. **Stem** is medium long with very little stem cavity. **Ripens** usually in early October. **Keeping** quality is one of the best.

Quality rating: **Good to Very good.**

◇ ◇ ◇ ◇ ◇ ◇ ◇ ◇ ◇ ◇ ◇ ◇ ◇ ◇ ◇ ◇ ◇

The Kandil Sinap apples on the tree in the picture are almost mature; even when small they are long and skinny. They are so beautiful and different that in summer I check them almost every day.

Kandil Sinap's history is shrouded in the last century. We know it was one of the Turks' favorite varieties around 1890. This means it is more than 100 years since it was first propagated. Apparently it is named after the Sinope Peninsula of Turkey, which juts into the Black Sea, and the name means the Sweet Apple of Sinope. It is still grown in the southern provinces of the former U.S.S.R. and in Turkey.

In my cooler climate I have trouble getting enough sugar in it without considerable thinning. It tends to drop badly if overcropped. It looks like no other apple I grow and attracts much attention at the amateur fruit shows.

Karmijn de Sonnaville

✳

Tree:

Synonym is Karmine™. **Origin** is from a cross of Jonathan x Cox's Orange planted in 1949 by P. de Sonnaville at the Wageningen Research Station, the Netherlands. Chromosomes are 3 x 17 = 51, a **triploid**. Tree is of moderate vigor and **size**. **Productivity** is average, but apples are larger and heavier than either Cox or Jonathan, the parents.

Annual bearer if thinned. **Winter hardiness** is unknown north of Zone 6 because of its newness in the U.S. I suggest trial in USDA zones 5 and 6 in the eastern U.S., and zones 7 and 8 in western Washington and Oregon. Summer climate should be on the cool side as in the Benelux countries, where it originated.

Disease susceptibility: in northern Europe to apple scab fungus, mildew and canker. Somewhat susceptible to scab on my two trees in western Oregon. **Blossom** season is the mid- to late-bloom. Do not rely on it as a pollinizer.

Sources known to me are Raintree, Northwoods, and Cloud Mountain Farm.

Fruit:

Size is medium to large. **Shape** is round to round-flat (oblate.) **Skin color** is similar to Cox Orange, but more striped and lighter. **Texture** is firm and somewhat coarse, with cream-colored flesh. **Flavor** is rich with a sharp, strong, sweet-tartness which mellows in storage. **Russets** often considerably in a cool wet spring, especially around the stem end. **Matures** from late September to early October. **Storage life** is adequate to February in ordinary cold storage at 30° to 32°F. **Physiological disorders** are some cracking on juvenile trees and blossoms are not cold hardy to frost. As a **triploid**, it has defective pollen.

Quality rating: **Very good.**

❖ ❖ ❖ ❖ ❖ ❖ ❖ ❖ ❖ ❖ ❖ ❖ ❖ ❖ ❖

I have an acquaintance who visits the U.S. every few years. He has a nursery and a commercial apple business about 65 miles from Paris, France. This man has excellent judgment, and once he commented that in his opinion the best tasting apple in Europe was a new one, not the old Cox Orange. He said this apple was a Cox Orange cross named Karmijn de Sonnaville, and some years ago he suggested the cultivar should be tested in the U.S., in Oregon, because the climate in our valley had some similarity to the Benelux countries and southern England. A statement like this to a man with an apple testing orchard is tantamount to waving a cape in the eyes of a snorting, sharp-horned, fighting bull. Instant action was required! Unfortunately, it takes time to process the budwood through quarantine and put them into production, but I now have two trees of the Karmijn de Sonnaville cultivar on M.9, doing well and fruiting.

The information below is quoted from notes sent by P. de Sonnaville, a fruit breeder at the Institute at Wageningen, The Netherlands. He has about 20 years of experience in fruiting Karmijn de Sonnaville.

Karmijn was planted in 1949 as a seed from a cross of Jonathan and Cox Orange, and is the only triploid known to come from a cross of these two varieties; consequently, it is quite self-sterile and does not pollinize other trees. Triploids generally set more fruit with two pollinizers, and in this case they need to be mid-season blooms such as James Grieve and many other varieties.

The flowers are large and beautiful and the fruit seems to grow best in a climate where the summer is not too hot or dry, and in moist air. The fruit is larger than Cox Orange, and, unfortunately, inherited the Cox Orange genes that make it subject to cracking and russeting. I suggest sprinkling with the hormone gibberellin after bloom to help prevent cracking and russeting. The flavor is outstanding and I think better than Cox Orange.

As with most triploids, the leaves are large and the branches are stiff and grow in a fine horizontal plane. The tree is only of moderate vigor and size, and it is moderately precocious (moderately early fruiting).

It is frost susceptible at bloom time, similar to Jonagold or Red Delicious. The flesh is juicy, hard and yellowish with both high sugars and good acidity. The skin color is yellow-green with carmine-red blush. The fruit shape is slightly broader than high (round to round-conical). Fruit is picked in The Netherlands from September 25th to October 3rd in most seasons, and stores well in cold storage to February or March. It is susceptible to scab and mildew.

What do I think about the statements of Mr. de Sonnaville? If you have had any success at all in growing Cox Orange, then my guess is that you will react as I did: you cannot wait until you obtain a tree of the cultivar. So much for addiction! The only cure is satisfaction.

When one bites into a Karmine there is such a reaction, it is difficult to keep from coming up out of your chair. Perhaps it is not for those who like mild, sweetish types of apples.

After observing many taste tests at two fruit shows, there is no doubt this is one of the best-liked of the high-flavored apples.

Keepsake

✳

Tree:

Origin was at the University of Minnesota in Minneapolis. Formerly MN.1593. **Parentage** was as a cross of MN.447 (open pollinated Malinda) x Northern Spy in 1936. Testing was started in 1947 and continued for 32 years before release to nurseries in 1979. Chromosomes are 2 x 17 = 34, a **diploid**.

Vigor and tree **size** are only moderate. It is usually an **annual bearer**. **Winter hardiness** is as good or better than McIntosh but apples **ripen** two to three weeks after McIntosh, which limits how far north it should be grown. I suggest lower USDA zone 4, zones 5 and 6, and also zones 7 and 8 in western Oregon. **Disease:** Keepsake is somewhat resistant to apple scab fungus, fireblight and cedar apple rust. **Blossom** season is mid-bloom, especially in the early part of mid-bloom.

Sources known to me: Fruit Testing Association Nursery, Inc., and Bailey Nurseries. Other northern nurseries may offer it by now.

Fruit:

Size is usually medium but if not thinned is often small. **Shape** is an irregular conic, often with angular sides—almost ribbed. **Color** on the sun-side is red with white dots, sometimes with scattered whitish scarf skin with a mottled look. **Stem** length is short. **Flavor** is quite aromatic when fresh, with some aroma being lost in storage. Fine dessert apple first two months after picking. It also ranks as a good pie and sauce apple after some **storage**. **Texture** is fine grained, crisp to hard, with light yellow flesh.

Matures in southern half of Minnesota between October 10 and 20, and about the same time here in the Willamette Valley. **Ripens** in 155 to 170 days from bloom. **Stores** well until March in less than ideal conditions, especially in slightly perforated plastic bags to prevent shriveling.

Quality rating: Good to Very good.

❖ ❖ ❖ ❖ ❖ ❖ ❖ ❖ ❖ ❖ ❖ ❖ ❖ ❖ ❖

Keepsake is one result of that fine apple breeder W. H. Alderman, from the University of Minnesota Horticultural Research Center, Excelsior, Minnesota, which is near Minneapolis. His crosses were responsible for such other north country apples as State Fair, and Sweet Sixteen. Most crosses were made in the 1930s and 1940s, with long test periods of 20 to 30 years before release.

The farm I grew up on was only a few miles south of the Experiment Station but, unfortunately, Keepsake was not known to the local farmers until many years after our farm was sold. We had to be happy with Haralson, Wealthy, Duchess, and certain crabs such as Whitney and Dolgo.

Today, if I lived on a farm again in central or southern Minnesota the first 5 apple cultivars I would plant (on cold-hardy rootstock—not Malling) would be Sweet Sixteen, Keepsake, Haralred, Honeygold, and Honeycrisp, all from the Minnesota breeding program. Among these different cultivars, harvesting would be from September 20 to October 20. We are in true north country now, with an occasional test winter of old-time severity. For culture in this north country, see the chapter under "Apple Orchard Culture Management."

Apparently Keepsake trees survive well in northern Minnesota, but ripen too late (about October 11 to 19) to be dependable. The University of Minnesota suggests growing it only in the southern half of Minnesota. It is not an exceptionally pretty apple, but I pay small attention to looks and have great respect for Keepsake's excellent keeping characteristic. The name is appropriate and its flavor is above average.

How it fares in quality, say in the upper south near Louisville, north Georgia, and other Mason-Dixon Line areas is unknown to me. In our cooler USDA zone 7 here in the Willamette Valley, it is still a winner. Because of my Minnesota heritage, I am about the only one who brings many of these northern cultivars to the fruit shows, and they compare quite well to top-tasting selections that grow in our milder climate, such a Jonagold, Braeburn, and Elstar.

Since Keepsake was not released for availability until the 1978-79 seasons, there is much left to learn about its adaptability to various climates. Its small to medium size may keep it from any large-scale commercial growing; however, it is an excellent back yard selection.

Kidd's Orange Red

✳

Tree:

Origin was at Greytown, Wairarapa, New Zealand. It was introduced in 1924 as a seedling cross of Red Delicious x Cox's Orange. Chromosomes are 2 x 17 = 34, a **diploid**.

A medium to small **size** tree. Slow grower. **Annual bearer** if thinned, but of only average productivity. **Winter hardiness** is probably not the best. I suggest growing from lower USDA Zone 5 through zones 8 or 9. **Disease susceptibility:** to European canker and somewhat susceptible to apple scab fungus. **Blossom** season is in the mid-bloom period.

Sources: only small or connoisseur type nurseries such as Southmeadow Fruit Gardens.

Fruit:

Size is medium to sometimes medium-large. **Shape** is conic and sometimes slightly ribbed. **Color** is scarlet stripes over a ground-color of yellow-green with patches of grey russet. **Stem** length is medium long, above the stem cavity and usually thick.

Matures from mid to late September. It hangs well so is often picked too late for storage. **Stores** for two months if not picked too late. Goes down in quality quickly after removing from storage.

Flavor and **texture** are above average with a sweet and juicy cream-colored to white flesh. Does not keep well under garage conditions but should be stored at 30º to 32ºF. I have noted no **physiological disorders** except some **russeting**.

Quality rating: Very good.

◆ ◆ ◆ ◆ ◆ ◆ ◆ ◆ ◆ ◆ ◆ ◆ ◆ ◆ ◆ ◆

Kidd's Orange Red was one result of the private fruit breeder, J. H. Kidd, in New Zealand in 1924. Ten years after this he gave us Gala, from a cross using Kidd's Orange Red, but it was the late Dr. Donald McKenzie who promoted Gala. J. H. Kidd also gave us the sweet, great-tasting Freyberg about 1939.

If Kidd's Orange Red would keep better after release from cold storage, it could have been a commercial apple despite some netting (broken russeting). In my orchard, most years the russeting is only light over the top two-thirds of the apple which gives it an unusual and, I think, attractive appearance. Commercial growers think russeting is detrimental to sales. In areas where frost is common at and after bloom time, the natural russeting is more severe.

I have fruited this apple now for at least 13 years and have always thought it of delightful flavor. Because it hangs so well in the tree, I originally picked it much too late for any kind of storage. It must be picked in my climate in mid-September for top quality. Pick slightly before Jonathan.

A delightful connoisseur apple for the discriminating grower.

Liberty

✳

Tree:

Originated as a known cross by the Geneva Station (Cornell University) in 1955 and introduced in 1978. **Parents** were Macoun x Purdue 54-12. Chromosomes are 2 x 17 = 34, a **diploid**. Tree **size** is medium. **Productivity** is so heavy the tree needs much thinning or apples are small to very small. **Annual** bearer. I suggest USDA zones 4 through 6 with trial in other zones. **Disease susceptibility:** it is immune to apple scab fungus in my area, and very resistant to cedar apple rust. Medium to good resistance to fireblight. Somewhat susceptible to powdery mildew fungus strain in the Willamette Valley of Oregon, but not a problem. **Bloom** time is from late early-bloom to mid-bloom. It has viable pollen and is heavily spurred. **Sources:** Fruit Testing Association Nursery, Inc., and other nurseries.

Fruit:

Size is variable, much dependent on culture and thinning amount. Sometimes medium, but often medium-small. **Shape** is usually round and somewhat oblate, but is sometimes conical, even on the same tree. **Skin color** is heavily blushed to sometimes striped a dark red over 90 percent, with ground color green to light yellow. **Stem** is usually very short, stout, and often barely visible. Not prone to **russet** except occasionally around the stem. **Matures** between October 1 to 10 at Geneva, New York, and Willamette Valley, Oregon. **Stores** well until December in regular cold storage of 30º to 32ºF. **Flesh** is white and aromatic; **texture** is very crisp; **flavor** is somewhat tart and refreshing.

Quality rating: **Very good.**

❖ ❖ ❖ ❖ ❖ ❖ ❖ ❖ ❖ ❖ ❖ ❖ ❖ ❖ ❖

Most of us who have worked with disease-resistant apples concede that Liberty has the best flavor of the scab immune cultivars; at least I think so. Goldrush may vie with it.

Dr. Robert Lamb of the Geneva Station in New York made the last cross. The previous crosses leading to Liberty included the flowering crab *Malus floribunda* which apparently supplied the dominant gene for disease resistance.

Only 16 years after its release by the Geneva Station, Liberty is being tested in small commercial direct market orchards in all apple growing areas. This rapid acceptance is attributable in part to the so-called "organic growers" who wish to use only limited, if any, chemical sprays to control disease. Home orchardists are, of course, also growing it. Liberty has unusual disease resistance combined with above average quality.

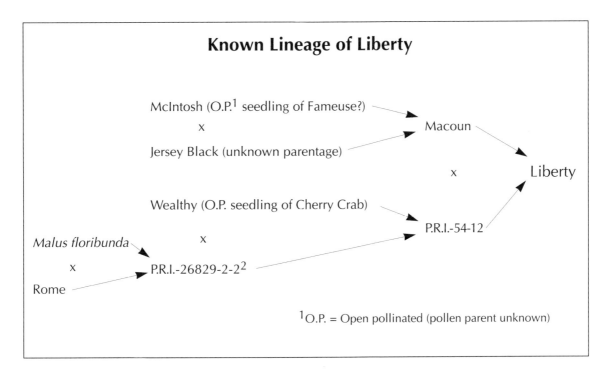

Known Lineage of Liberty

McIntosh (O.P.[1] seedling of Fameuse?)
x
Jersey Black (unknown parentage)
→ Macoun

Wealthy (O.P. seedling of Cherry Crab)
x
Malus floribunda
x
Rome
→ P.R.I.-26829-2-2[2]
→ P.R.I.-54-12

Macoun x P.R.I.-54-12 → Liberty

[1]O.P. = Open pollinated (pollen parent unknown)

The following apples have had McIntosh as one parent, and all strongly resemble this parent in color. They are Empire, Jonamac, Macoun, and Spartan. Liberty is only part McIntosh but varies so much in shape it can resemble all of them. This is a challenge to proper identification.

Liberty grows to perfection in USDA zones 7 or 8 in Oregon. How it will perform in humid east Texas, also zones 7 and 8, is unknown to me. It appears to do well from Wisconsin to the Appalachians of north Georgia, in cooler climates.

It is, of course, not resistant to insect damage but for those wanting to hold spraying to the absolute minimum, Liberty is an excellent choice. When the first few ripe apples drop, they must all be picked immediately, or two-thirds of the crop can drop in a few days.

Highly recommended!

McIntosh

✳

Tree:

    Synonym: before 1835, it was called Gem. Its **parentage** is unknown, but it is thought to be from the very old Fameuse, which it much resembles, especially in that both have snow-white flesh. Chromosomes are 2 x 17 = 34, a **diploid**.

    In **size** it is a moderate to fairly large tree. It tends to be somewhat **biennial** in fruiting habit, but has some apples in the off year. **Hardiness** is above average and it does very well in USDA zones 4 and 5, with some of zone 3 if protected. **Disease susceptibility:** to apple scab fungus. **Blooms** from the late-early to mid-bloom period.

    Sources: most commercial nurseries, with many redder strains available.

Fruit:

    Size is medium to medium-large. **Shape** is round to round-flat. **Skin color** is bright red and sometimes striped carmine, often dark purple-red. The **stem** is usually short and of medium thickness, sometimes knobbed. It is highly **flavored** and rather acidic. **Flesh** has perfumey aroma, even before being cut open. **Texture** is rather soft and becomes more so after two months of storage. **Stores** well in C.A. but for a rather short time in common storage at 30° to 32°F.

    Matures in mid to late September, except in its northern limits (USDA zones 3 and 4) where it ripens in early October. In zones 5, 6, or 7, it tends to **drop** 20 to 30 percent of its crop before ripe. There is some dropping even in the far north. Its **physiological disorder** is uneven ripening and dropping.

Quality rating: **Very good to Best** when grown in USDA zone 4, and perhaps some of zone 3.

✧ ✧ ✧ ✧ ✧ ✧ ✧ ✧ ✧ ✧ ✧ ✧ ✧ ✧ ✧ ✧

    Years ago I heard rumors that the McIntosh was really quite old, maybe dating from around 1795 or so. It seemed only appropriate to check reliable old sources for confirmation.

Beach, in *The Apples of New York*, concentrated on one fruit (apples) and went to more length to cover the history of an important apple cultivar. He rates McIntosh "Very good to best" in quality, and writes in 1903 that it is one of the most promising varieties! Beach was somewhat ambiguous about the correct dates, saying: "It was a chance seedling found in Dundas County, Ontario, where Allan McIntosh began the propagation in the nursery about 1870." The date is quite inaccurate. I can only speculate as to why a great horticulturist did not know the true history or did not think it important enough to investigate carefully.

In 1990, McIntosh was first in production in Canada and third in the U.S., so Beach's prediction of "promising" came true, but primarily due to two subsequent weather occurrences he could not have foreseen. Beach wrote at that time that Baldwin was the number one cultivar in the northeastern United States, followed by Rhode Island Greening and Northern Spy.

In the winter of 1918-19, just after World War I ended, the northeast had a severe cold spell that damaged or killed outright a great many of the Baldwins, which are not extremely hardy. Then, 14 to 15 years later, in the winter of 1933-34, an even worse cold spell in the northeastern U.S. destroyed the Baldwin almost completely and, of course, killed a certain number of Rhode Island Greenings and Northern Spys.

By the time of these disasters, the much more hardy McIntosh had been slowly gaining favor, at least in Canada; however, these two severe winters gave the McIntosh a quantum leap, as it alone almost completely filled the big hole created by thousands of dead or dying trees. Ultimately, the Baldwin tendency to a biennial or even triennial bearing habit probably would have driven it to a very low production, once land became more valuable after 1945.

So, what is the accurate history of the McIntosh apple and the McIntosh family? We find an answer in a book titled *North American Apples: Varieties, Rootstock, Outlook*, published in 1970 by Michigan State University Press, East Lansing, Michigan. A number of authors and horticulturists contributed in the 1960s and, nearly 30 years later, the production statistics are out of date but the interesting history of McIntosh is not. This book is still available from Michigan State University Press.

For the history up to about 1875, we will use information in part from W. H. Upshall's chapter on McIntosh. (He was the former horticultural director of the Experiment Station in Vineland, Ontario.)

The history of the McIntosh apple is intimately tied to the McIntosh family for three to four generations. Alexander McIntosh emigrated about 1776 from the Scottish Highlands to the Mohawk Valley of New York State. Their youngest son John was born there in 1777. When he was 19 years old, he had a disagreement with his parents over a young lady and left home for Canada in 1796. There he met and married another young woman named Hannah Doran in 1801, and farmed for 10 years along the St. Lawrence River. In 1811 he traded farms with his brother-in-law, Edward Doran. On this new farm, several miles from the river, about ¼ acre had been cleared some years before 1811. Among this brush were several apple trees that must have been only a few years old, for they were dug up and moved to a garden area. Originally, they were probably planted between 1803 and 1807. No one is sure.

Apparently McIntosh fruited about 1810 to 1812 or so, and for the next few years John McIntosh thought it such a good apple they planted many seeds of it, none of which gave another McIntosh, although some were similar. The problem was that none of the McIntosh family or their neighbors knew how to graft or bud.

About 1835 an unnamed itinerant worker came through the area and he was skilled in grafting and budding. He taught the McIntosh family how to do it. The family had started a nursery in 1820 selling seedling trees, but now in 1836 they could start selling real McIntosh grafted trees that would give true McIntosh apples.

John McIntosh's farm operation expanded and his wife Hannah took on more responsibility for the orchard and nursery of their soon-to-be renowned apple. At this time, it was known as Gem and Granny's Apple, but by 1836 it was named McIntosh Red. Eventually, it was called just McIntosh, after the family name.

It was at about this time (1836-37) that son Allan began to take over the nursery business. Allan had used McIntosh seedling rootstock to graft onto, but soon shifted to the more hardy Transcendent Crab rootstock. Allan was aware that pollinizers were needed and sold other varieties as well, to increase fruit set.

He also had setbacks in his nursery, the worst in 1866, when his cattle broke through his fences and ate or trampled about 10,000 small grafted McIntosh trees that would have been sold the next spring. Allan's younger brother by 10 years, Sandy, was an expert grafter and salesman and did much to publicize their apple by offering taste tests and selling trees.

Allan's youngest son, Harvey (1863-1940), then took over the business in the late 1800s from 50 acres of inherited land, plus a good deal of leased land. In late winter 4 or 5 men were hired to make bench grafts for planting in spring as there was by now great demand for McIntosh trees.

Harvey's only son, another Allan, took over the nursery in 1940, but not too many years ago the nursery business became secondary to 75 acres of apples, mostly McIntosh.

Unfortunately, the McIntosh home was built only 15 feet from the original tree and, when the house burned down in 1894, only desperate efforts and later repair work saved it from instant demise. It last fruited in 1908, but it had been so badly damaged by the fire it fell over in 1910 at 100 years of age or more. One wonders how long it could have lived.

I have a strong curiosity about the real history of exceptional apples, old or new, and hope the reader enjoys finding such information in this book. We are indebted to W. H. Upshall's research on the McIntosh.

McIntosh does best as far north as it can be grown where it is a late fall or winter apple with more carbohydrate solids, better flavor and keeping qualities. The best areas are around its home in the upper St. Lawrence Valley and in parts of Quebec, Nova Scotia, northern Michigan and upper New England.

I have eaten many McIntosh apples from the Okanagon area of British Columbia and find them fine just off the tree but, when purchased in Portland, Oregon, in February and March, only 350 miles from the Okanagon, they are usually soft and bruised. I think it a rather finicky apple to grow to good quality, except under the best condition of long daylight hours that are not too hot and cool nights from September 1 to ripening.

Farther south than the above-mentioned areas, it can become a fall apple that drops badly, is softer and not as good in flavor. At its best it is perfumey, of unusual flavor, and does well in areas too cold in winter for many worthwhile cultivars to survive.

On seedling rootstock it is of moderate vigor to semi-vigorous but, in the far north, the more severe climate no doubt holds the size down. Some of the cold hardy seedling types used now for rootstock (Borowinka, Beautiful Arcade, Dolgo, etc.) may be 10 to 25 percent dwarfing, and in my estimation should be used instead of seeds from canneries, which may be diluted genetically with less hardy types such as Red or Golden Delicious.

Melrose

✳

Tree:

Crossed about 1931. **Parentage** is Red Delicious x Jonathan. Chromosomes are 2 x 17 = 34, a **diploid**. Tree is a vigorous grower and it is quite large in **size** even on M.7A rootstock, which my oldest tree is on. **Annual** bearer if apples are thinned shortly after forming. **Winter hardiness** is best in USDA zones 5-8. **Disease susceptibility:** to apple scab fungus and slightly susceptible to powdery mildew and anthracnose. **Blossom** season is mid- to late-bloom.

Sources: most nurseries selling apple trees.

Fruit:

Size is medium to quite large. **Shape** is variable on the same tree; some are quite flat (oblate), others are more conic like the Red Delicious parent. **Skin color** is red over yellow-green. In hot climates, the red is frequently rather dull and russeted around the stem. In Skagit county in northwest Washington (north of Seattle) it frequently **russets** on one-half the apple or more, probably due to cool, wet spring weather. In our area we usually get a red apple.

Stem length is rather short in a shallow stem cavity, but they usually come to the top of the cavity. They should be thinned to singles because short stem clumps of two or three fruits may push some off, especially if apples are quite large. **Texture** is breaking and very juicy. **Flesh** color is white. **Flavor** is above average if apples are well colored with no excess use of nitrogen after the first three or four years of growth. **Maturity** is October 10-25 in most zones.

Storage life is above average, to March or later at 30° to 32°F. It even keeps well under cool garage conditions. Melrose seldom has **physiological disorders** except for **russeting** and, in large apples, some bitterpit. (See chapter on bitterpit.)

Quality rating: **Very good.**

❖ ❖ ❖ ❖ ❖ ❖ ❖ ❖ ❖ ❖ ❖ ❖ ❖ ❖ ❖ ❖

Melrose is an example of modern breeding, done by Freeman S. Howlett of the Ohio Agricultural Experiment Station in Wooster, Ohio, around 1931. It was selected in 1937 and released in 1944. Harvest time is about 14 to 20 days after Red Delicious, and about three weeks after Jonathan. It seems adaptable and gives an above-average apple in many fruit growing areas; however, primarily because of their climate, the Yakima Valley in eastern Washington has properly ignored it. Their excessive summer heat tends to ripen it too soon and it does not seem to color there as mine do. I am quite certain this lack of color in that area is also due to using too much nitrogen. It is an apple for the cooler, mild fall areas with proper fertilizing management.

One of my trees is over 22 years old, and I have gradually moved picking time from about October 10 to October 20 to 25, as it is one of those rare apples that seem to keep just as well picked fairly ripe and tasty as when picked too early and starchy.

This is one apple that needs red color from sunlight; not 100°, but 70° to 85°F in September and even into early October, plus some cool nights (35° to 45°F in late September/early October.) Prune the branch whorls far enough apart for sun entry. It also must not be overcropped or there will not be enough leaf surface to form the carbohydrate solids (sugars); Melrose is often short of those. If you do not follow this program, you may get green-colored apples and then Melrose can go from "very good" in quality to "Blah!"

Except for Ohio, the Willamette Valley of Oregon planted Melrose before most areas because the fine pomologist Dr. Quentin Zielinski, who moved from the Midwest to Oregon State University about 1949, apparently brought grafting wood he gave to his relatives, the extended Zielinski family around Salem, Oregon, some of whom are commercial growers. It has been grown for 40 or more years, especially in back yards in our valley. Every direct market grower I know from Portland to Eugene, Oregon, sells at least some Melrose apples.

My trees are almost certainly only second generation from the mother tree. One researcher says the most perfect Melrose color and taste he has seen are selections from this valley which appear at our fruit shows. If you are not too far north (such as zone 4), this apple should be in your orchard. One descriptive sentence sums it up: Melrose is my beautiful, annually dependable, fine-tasting, multiple usage, long-keeping apple—a family favorite.

Mutsu

*

Tree:

Origin: called Crispin in England and New York state, it was crossed about 1930 at the Aomori Station in Japan. Its **parents** were Golden Delicious x Indo. Indo is a large, green, late-ripening apple, no longer commercial in Japan. It was frequently used for breeding purposes. Named Mutsu in 1948 and patented in 1949. By the late 1960s, it was tested in many areas of the world and was added to my test orchard in 1973.

Tree is a **triploid** with 51 chromosomes and has virtually no fertility for itself or other apples. Do not rely on it as a pollinizer. **Blossoms:** early to mid-season bloom. **Winter hardy** through much of zone 5. Grows to perfection in Western Oregon and Washington in zones 7 and 8 with our cooler but long season. **Biennial bearing** with average culture. Early chemical blossom thinning helps to correct it. Tree **size** is large and vigorous. **Disease susceptibility:** to apple scab fungus and powdery mildew.

Sources: most small and some large commercial nurseries.

Fruit:

Size is large to sometimes very large (up to 1 pound each). **Shape** is conical to oblong (irregular), often ribbed and angular. **Skin color** is green to sometimes a light yellow when picked late. **Stem** is usually long and slender but is variable in length. **Flavor** is a balance of sugar and acid, and a coarse **textured**, firm fruit. **Russet** shows occasionally at either end, but not usually.

Storage life is long if the apples do not have bitterpit. (I suggest five to seven calcium chloride sprays during the growing season and low nitrogen to prevent bitterpit.)

Quality rating: **Very good.**

❖ ❖ ❖ ❖ ❖ ❖ ❖ ❖ ❖ ❖ ❖ ❖ ❖ ❖ ❖ ❖

Mutsu seems to be having trouble finding a niche in the American apple market, but it has already become a favorite of many direct market growers where the climate is not too hot, especially here in the Willamette Valley of Oregon, where it is bringing high prices.

In the hot central valley of California, with high temperatures and large-size fruit, bitterpit often destroys the Mutsu market. On the other hand, in the higher, cooler foothills at Miramonte, California, one grower says Mutsu attracts more repeat customers every year due to its good size and excellent flavor.

Here in the Willamette Valley we pick Mutsu in late October, but I have found this cultivar to be an erratic bloomer, and it sometimes sets a poor crop. It was an outstanding keeper in common storage. It does not shrivel in storage, and it is very resistant to spray damage.

In a trellis line, a Mutsu tree should probably be spaced about four feet wider than a tree with a moderate growth rate because of its vigor and size.

I like this apple very much for fresh dessert. It also makes above average pie, sauce, and apple juice mix. Mutsu is an all purpose apple when it is grown with good management and in a climate without extremes. It is often reported to be difficult to grow, but I do not find it so in our climate.

N.Y. #429

✳

Tree:

Origin is at the Geneva Experiment Station in Geneva, New York, a division of Cornell University. **Parentage** is Red Spy x Empire, and chromosomes are 2 x 17 = 34, a **diploid**.

Tree is of average **size** and vigor. It has many fruit buds and tends to overset as does Empire, one of its parents. In my orchard it has been an **annual** bearer. **Winter hardiness** is not well known to me because it is not yet widely grown. I suggest USDA zones 5 through 8, with trial in protected zone 4. It is **disease susceptible** to apple scab fungus. **Blossom** season is in the late mid-bloom, with Melrose, Red Delicious, and Golden Delicious.

Source: Fruit Testing Association Nursery, Inc. This is the only source as of this writing.

Fruit:

Size is medium to large. **Shape** is round, similar to Empire, but larger. **Skin color** is very beautiful, a dark pinkish-purple when ripe. **Stem** length is medium. **Texture** is crisp and breaking. **Flavor** when ripe is rich and different, a cocktail of flavors; it scores high in taste tests.

Matures in zone 7 of the Willamette Valley from October 5 to 20. Will hang on the tree longer. **Storage life** is at least two to three months in common storage at 30° to 32°F. No **physiological disorders** noticed.

Quality rating: **Very good.**

❖ ❖ ❖ ❖ ❖ ❖ ❖ ❖ ❖ ❖ ❖ ❖ ❖ ❖ ❖

N.Y. #429 has now fruited in my orchard for eight years, and each year it gets more respect and attention as an apple to be thoroughly tested. Those of us who buy these numbered apples from the Geneva Station are bound by the testing agreement to report our results, which I have done.

As grown here, most anyone would agree it is a beautiful apple. It is nearly as dark as Arkansas Black but with dark pink overtones blended with the reddish black.

I think Empire, one of N.Y. #429's parents, is perhaps the best of the McIntosh crosses. Most of the others have too many of McIntosh's faults—such as soft fruit, pre-harvest drop and poor keeping. The other parent, Red Spy, imparted some of Spy's good keeping and fine flavor. It is the only apple that has not yet been named that is receiving high marks from many testers.

Similar to Empire, when #429 is bitten into it responds with a chunk of apple (breaking) and a crack which is audible, at least to the apple eater. We should pay more attention to this one. Yes, it is similar to #428, which is a Scholarie Spy x Empire cross but, in my opinion, #428 does not have quite the color or flavor of #429.

This apple has been numbered for some years, but I think it will soon be named and perhaps patented. We understand there are other similar cultivars, and Geneva is debating which to name first. By the time this book is published, perhaps the Geneva Station will have given #429 a name. If so, I believe this will help make it better known. After all, it is somewhat like bragging about one's only grandchild as #1, instead of "Rachel," which we really prefer and which gives her real identity.

Newtown

✳

Tree:

Synonyms: Newtown Pippin, Yellow Newtown, Albemarle Pippin. The tree is quite pyriform when young, and I advised using branch spreaders the first few years. It is probably **not very hardy** and grows best in favored areas of USDA zones 5 through 8. Newtown **bears** one to two years later than many apples on size-controlling rootstock and may take many years on seedling rootstock (which I do not recommend.)

A **diploid** of 34 chromosomes, it is slightly **biennial** if not thinned but seems somewhat self-fruitful. **Susceptible** to apple scab fungus and powdery mildew.

Sources: widely available from western nurseries; spur strain is from Carlton Plants and Brandt's Fruit Trees, Inc.

Fruit:

Apple **size** is small-medium to sometimes medium-large. **Shape** is round-conic to often flattened. **Skin** is green, with russet rays around the stem on about 60 percent of the fruit. **Stem** is medium to short. **Flavor** is unusual, piney and delightful, and excellent for all uses except salad, as the flesh browns almost immediately when cut.

Storage life is long, even under garage conditions. **Picking time** here is late October, but I have picked them in late November when it was cool and wet. Pick earlier in the southeast (North Carolina, the Virginias.)

I hope readers will try to grow this apple in the USDA zones where it can succeed. It could be another 100 years or more before someone again finds an apple that can be rated as highly as Newtown is by many of us.

Quality rating: **Best** (the only one in this book with this rating.)

❖ ❖ ❖ ❖ ❖ ❖ ❖ ❖ ❖ ❖ ❖ ❖ ❖ ❖ ❖ ❖

Newtown was reported to have been found as a chance seedling in the early 1700s on the estate of Gershom Moore, near the village of Newtown on New York's Long Island. Thomas Jefferson's notes reported he planted about 170 Newtown trees over a period of 44 years. By this sheer quantity, one might assume they were one of his favorite apples.

Today, Newtown grows to perfection in certain areas of California and Oregon. About four million boxes a year are still grown commercially, primarily in those two states. It is still in the top 15 cultivars. (See USDA chart.)

I first tasted well matured Newtowns while studying plant pathology and horticulture at the University of Minnesota many years ago, but we could not ripen them in zone 4 of southern Minnesota and had to import them from Oregon. It immediately became a favorite of mine, and will always remain so. It is the only apple cultivar that I have on four different rootstocks. They are quite disease free in my climate if sprayed for apple scab fungus. One 22-year-old tree on MM.106 gives us about 5 bushels of apples each year. Because of more tannins than most commercial apples, it is especially valuable in a sweet cider mix, but these apples are so desirable for other uses it seems a shame to use them in apple juice. Newtown is especially liked for fresh dessert but is equally sought after for pies, sauce and drying.

Some of the probable history in Virginia surrounding this apple is most interesting. General Braddock was the British Commanding General of the forces against the French in 1755. On July 9, General Braddock personally led one of his three armies, of about 1,600 men and 89 officers, against Fort Duquesne in eastern Pennsylvania. A few hours later in the battle, the General had only about one-half his force left, and he was so severely wounded he died. There was a militia from Virginia involved in this battle which was commanded by a 23-year-old officer in the British Army. With only about 30 men left, this Virginia officer decided to retreat. He learned that one can retreat to fight another day, as this man did 20 years later, but this time against the British. His name was George Washington.

Fleeing with Washington was Captain Thomas Walker, M.D., of Castle Hill in Albemarle County, Virginia. According to Beach in *The Apples of New York*, Captain Walker had previously married the widow of Nicholas Merriwether, thereby coming into possession of the estate called Castle Hill, a few miles from Monticello.

Captain Walker was the Commissary Officer under Braddock, and probably also acted as physician to these troops. The remainder of this defeated army went into winter quarters. Captain Walker, at that time or in the fall, apparently filled his saddle bags with apple cuttings, some no doubt from Newtown trees in the vicinity of Philadelphia. On return to his home at Castle Hill, the cuttings were used in his orchard, as well as being distributed in his immediate vicinity in Albemarle County (Beach).

Starting later, perhaps about 1850, these apples were thought to be slightly different from Newtown, perhaps a seedling, and were then called the Albemarle Pippin. It was named after the county of Albemarle, where Castle Hill and Monticello are located.

U. P. Hedrick followed Beach at the Geneva New York Experiment Station in the early 1900s. It was his opinion that there are no strains of Newtown. He said Newtown apples vary more than most in different climates and soils, which had led to the belief there were green, yellow and Albemarle Newtowns but, when grown side by side, they were the same.

In 1984, about 250 years after Newtown was found, a spur strain was finally discovered in Hood River, Oregon. I now have one of these multiple spur trees, a four-year-old, growing in my orchard, and I look forward to checking its tree size on MM.111 rootstock.

Apple growers in the Hood River, Oregon, area grow more Newtowns than other local apples, but they appear concerned about consumption because of Granny Smith imports. In recent years Newtowns have been picked early with no pink or purple color showing (three to five weeks early in my opinion), so they could compete with the green, tart Granny Smith imported from New Zealand. These Newtowns then are hard, rather tasteless rocks that will frequently get bitterpit in storage. This may have turned off some Newtown lovers, but in my estimation it is far superior to Granny Smith for all uses—when picked ripe! Indeed, Newtowns picked at the correct time are not nearly so subject to bitterpit. They keep better when picked in late October here in Oregon. Fortunately, even the ones that are picked early ripen somewhat in storage with an increase in sugar, but they never catch up to the flavor of those picked at the proper time.

Unfortunately, Newtown is more difficult to grow then Granny Smith, but with proper management a commercial grower can still make money with it, especially if he direct markets and offers taste tests. I don't have any packers or sales groups dictating to me and I think my trees are properly managed. The apples have no bitterpit, and my family and friends enjoy them from the time they are picked until April of the following year.

Northern Spy

✳

Tree:

Synonym: Spy. **Originated** in New York about 1800. **Parentage** is unknown and was a chance seedling on the farm of Heman Chapin. Chromosomes are 2 x 17 = 34, a **diploid**, but is quite a late bloomer and needs a late-blooming pollinizer. Vigorous grower, large in **size** with pyriform shape unless trained. (Tends to grow almost straight up.) Productivity is medium and tends to be **biennial**. **Winter hardiness** is from lower USDA zone 4 and does best in zones 4, 5, and 6. Seldom freezes out in a test winter in zone 5. **Disease susceptibility:** to apple scab fungus and sometimes in storage to a blue mold in the Penicillium group.

Sources: most commercial nurseries. Some offer a strain called Red Spy. My experience with this redder strain (or mutation) is that it ripens one to two weeks earlier and does not keep nearly as well, so I do not recommend it.

Fruit:

Size is medium-large to large. **Shape** is usually conic, but sometimes is round-conic and usually slightly uniformly ribbed. **Color** when exposed to the sun is red and red stripes over yellow ground color. Ground color can be green if not properly ripened or sun-exposed. **Stem** is medium to long, usually moderately thick and usually slightly above the cavity. **Skin** is tough, thin and smooth. **Flavor** is sharp and high flavored. **Flesh** is somewhat yellowish; **texture** is fine, tender, juicy, and crisp. Seldom if ever **russets**. **Pick** about October 15 to 30 in the Willamette Valley climate, when well colored. With excess nitrogen the apple loses flavor and does not color well. **Keeps well** in common cold storage at 30° to 32°F. **Physiological problem:** is susceptible to bitterpit.

Quality rating: **Very good to Best.**

◇ ◇ ◇ ◇ ◇ ◇ ◇ ◇ ◇ ◇ ◇ ◇ ◇ ◇ ◇

According to Beach, Oliver Chapin and his brother Dr. Daniel Chapin traveled from Connecticut to East Bloomfield, New York, in 1790. They brought apple seeds with them and planted them. Ten years later, in 1800, their brother

Heman arrived, who purchased 200 acres west of his brothers' acreage, and also planted apple seeds. The original Northern Spy tree then first appeared on Heman's land; however, this original young tree died before fruiting. Northern Spy was rescued from oblivion because Heman's brother-in-law, R. Humphrey, had planted some suckers from this seedling tree before it died.

In this way, a rare new apple was born. I describe it as "rare" because not many apple cultivars are in the quality category of "Very good to Best" for fresh eating, are outstanding for processing and also keep well. Northern Spy has all three of these qualities. Poor tree characteristics and bitterpit susceptibility keep it from greater production. It does best in a line just north of Chicago straight east. For perfection it must ripen as a late fall, early winter cultivar.

By 1840, many orchardists knew about Northern Spy, information about it having been spread largely by word of mouth. In 1852 the American Pomological Society recommended it for general planting in other areas (Beach.)

Horticulturists frequently state that this apple tree does not have its first apple crop until 10 to 15 years after it is planted. This is only true if Northern Spy is on certain seedling rootstocks. In 1903, Beach wrote that under favorable conditions this cultivar often gave a profitable crop in 7 years (on seedling rootstock). Unfortunately, he did not specify what he meant by "favorable conditions."

My first two trees of Spy on M.7A rootstock fruited in the fourth year and had nothing the next year, reflecting its biennial habit. On M.9 rootstock they usually fruit in the third year and are nearly annual in habit. M.9 rootstock helps correct biennial bearing in many cultivars such as Northern Spy and Criterion. The reason is unknown. In the cold areas of USDA zones 3 and 4, Columbia Crab, Beautiful Arcade or other cold hardy rootstock should be used, perhaps with a cold-hardy interstem from the Polish or Budagovsky series, to reduce the size of this big tree.

The Northern Spy tree is not without other faults, and that is why I prefer using a trellis when growing it. The branch wood is very straight grained and breaks rather easily, as does Gala, when carrying a load of apples (especially when the tree is young.) If Northern Spy is grown as a free-standing tree on, say, M.7A rootstock, one would need to use limb spreaders from day one because of the pear-like (pyriform) habit its branches have of growing nearly straight up.

Years ago, Spy apples were often not pollinized because the tree bloomed so late, and it would give small, seedless apples of poor color, size and quality. Today, we have flowering crab apple trees that bloom late, and some can pollinize Northern Spy. Other late bloomers such as York, Orleans or Wealthy can be used. Golden Delicious also blooms during the early part of Northern Spy's bloom, but Golden is not cold hardy at Spy's northern limit.

If grown so that the apple colors properly, Northern Spy keeps for an above-average length of time. The apple has a thin but tough skin and rather tender flesh, and does not take shipping (even in tray lined boxes) as well as some of the other top cultivars. It is not as tender as McIntosh when being shipped, but it is close.

Northern Spy is widely recognized as being an excellent processing apple that is outstanding for apple pie slices, apple sauce, and apple juice mix. Some do not like this apple when it is baked, but I do. It has a sharp distinctive flavor but does have that tough skin. Because the apple is so juicy and tender, it is only fair for drying.

Northern Spy was still in the top 15 U.S. commercial cultivars in 1989. It seems to be gaining slightly in Canada, and is holding its own in the eastern U.S. Except in backyards, the western U.S. grows very few Northern Spy trees.

The great Armistice Day storm of 1940 came early, on November 11, and killed 89 percent of all Northern Spy trees in Iowa (as reported by Iowa extension people). One hundred percent of the Red Delicious trees were killed. In contrast, all the Duchess of Oldenburg trees survived at our farm near Northfield, Minnesota, as did all Wealthy, Haralson and the crabs.

The severe winter of 1933-34 in the northeast U.S. was especially hard on Baldwin, Rhode Island Greening and Northern Spy trees. Northern Spy trees that were located near the coast survived, but many of those inland did not. Dry, dehydrating low temperature winds aid in causing the demise of many cultivars in areas that experience those occasional severe test winters. Northern Spy continues to grow into the fall and has a late leaf fall. It is often not in full dormancy when a severe mid- to late-November freeze occurs.

Some areas of zone 4 may be too cold for Northern Spy. St. Lawrence county is actually in zone 3 in New York in a flat, open area northwest of the Adirondack Mountains, too far from the St. Lawrence River to experience the river's buffering effect, and it often fails there. This area, along with the upper regions of Vermont, Maine and New Hampshire, are usually not good places to plant Northern Spy, especially at high altitudes. Every few years these areas have winter temperatures of minus 30º to 40ºF with dehydrating winds.

I suggest growing Northern Spy on the rootstocks in the Trellis Chart to keep the tree to manageable size, but in zone 4 put this cultivar on one of the cold-hardy rootstocks. It will usually not survive a test winter in zone 3 unless on cold hardy rootstock and, even then, it may not live.

Orleans

✳

Tree:

Synonyms: Orleans Reinette and Winter Ribston. Probably of French **origin**. **Parentage** is not known. A **diploid** of 34 chromosomes. **Production** is average to above average, and is **annually bearing** if thinned early. Tree is upright growing. **Blooms** quite late in western Oregon, with, or later than, Northern Spy.

Winter hardiness is unknown, but I suggest USDA Zones 6 through 8, with trial in lower Zone 5. **Disease:** fairly resistant to apple scab fungus; somewhat susceptible to powdery mildew.

Southmeadow Fruit Gardens is one of the few **sources** of this fine connoisseur apple.

Fruit:

Size is usually medium, but occasionally is medium-large, and quite uniform. **Shape** is flat-round with a wide shallow calyx basin. **Skin color** is reddish, mixed with yellowish fine russet. **Stem** is short to medium and usually thick. **Flesh texture** is somewhat coarse and cream colored to white. **Picking time** is usually mid-October. **Storage** is usually not more than two months and shriveling is kept at a minimum by putting it in plastic bags with a few pinholes. **Flavor** is rich and unusual.

Quality rating: **Very good.**

❖ ❖ ❖ ❖ ❖ ❖ ❖ ❖ ❖ ❖ ❖ ❖ ❖ ❖ ❖ ❖

This is a very old apple, and how and when it got into England is not known to me. It has a very French name, but no one seems certain it is of French origin, but it probably is. In the past, England has given it certificates of merit, and it was first mentioned in the late 1700s.

The very old term of Reinette was probably attached to one particular cultivar somewhere between 1400 and 1650 A.D., but should probably no longer be used as it cannot be properly explained. Orleans is a fine name and seems better without a confusing nickname attached to it, although that is only my opinion; some prefer the nickname attached.

My first tree was on M.7A rootstock, purchased some years ago from South-meadow Fruit Gardens in Lakeside, Michigan. Originally it was planted in clay loam, somewhat on the light side, and given no nitrogen. The first two crops gave apples of exquisite flavor, with a considerable red color over some slight russeting. The tree was then moved to a shadier part of the orchard in heavier clay with compost and nitrogen mixed as a top dressing. Subsequent crops have not been as high in quality or color, so this suggests some nitrogen sensitivity. I now have another two trees on M.26 rootstock in lighter soil and will try to get that original quality back by applying only small amounts of nitrogen after the first two years or so. One does not seem to live long enough to ever finish testing. Testing is more of a journey than a destination, especially with very old cultivars on different rootstocks and in different soils that few others have. Very little comparison is available from others.

This is one of the better antique apples that is not well known in America. It is late maturing, and England would probably have to grow it in a warm micro-climate in the south to get quality. Unless red-cheeked, it has considerable resemblance to the more ancient Court Pendu Plat but tastes much better, with a finer texture. Perhaps it is a seedling of the Court, which may go back to Roman days, as both are quite late blooming. It also resembles the flat Blenheim Orange.

A few years ago, a garden writer wrote in one of London's daily papers that Bramley is their best cooking apple and Cox Orange is a terrific eater, but Orleans is something special and much overlooked. I would add that my above explanations may mean that climate, soil, and management play a great part in its quality.

It has a somewhat irregular bearing habit and, if the tree is not thinned well and overcrops there will be some, perhaps considerable, fruit drop. This is a very rare apple here in the west, so I can only go by my own 14 years of experience with it. Its scab resistance and good flavor make it an above-average connoisseur apple.

Rhode Island Greening

✳

Tree:

Its exact **origin** (probably Green's End, Rhode Island) and its **parentage** are unknown. Chromosome count is 3 x 17 = 51, a **triploid**. It was probably planted from seed about 1650.

A long-lived tree that eventually becomes quite **large** on seedling rootstock. Like many **triploids**, it tends to bear one year later than most apples. A **biennial** bearer unless grown on the lighter soils, and limed and fertilized as it grows older. **Hardiness:** can be grown in certain parts of USDA zone 5 in New York, but zone 6 near water, especially, is its best habitat in the eastern U.S. In the cooler areas of western Oregon and Washington, zones 7 and 8, I know of two trees well over 100 years of age. Zones 7 and 8 in the eastern U.S. are probably too warm inland, which makes it a late summer apple of lesser quality.

It is quite **disease susceptible** to apple scab fungus and certain cankers. **Blossom season** is the early part of the mid-bloom period. Do not use King, Gravenstein, Stayman, or Baldwin, as pollinizers because all five are triploids.

Sources: eastern nurseries and Southmeadow Fruit Gardens in Lakeside, Michigan. This is surely the oldest cultivar of American origin in the U.S. that has been in continuous commercial production. Roxbury Russet is thought to be 20 to 40 years older, but it is no longer grown commercially except for direct marketing in some communities and by home orchardists.

Fruit:

Size is medium to large. **Shape** is variable. It is usually round or round-flat (oblate), but some of the fruit may be elliptical like some Roxbury Russets. **Skin color** is usually green, sometimes with a reddish blush. If allowed to get yellow on the tree, it will not keep well, but it has a peculiar and desirable flavor. It should be **picked** green.

Stem is medium to short and medium thickness, often with some pubescence (hairiness.) **Skin** is fairly thick, tough, and waxy. It **russets** a little at times. **Flavor** is rich, somewhat tart, with a very different taste. **Flesh** is somewhat yellow, firm and juicy.

131

Maturity is early October in the northeast U.S. but usually somewhat later in western Oregon and Washington. **Storage life** is about December to January in regular cold storage of 30° to 32°F.

Quality rating: **Very good**.

❖ ❖ ❖ ❖ ❖ ❖ ❖ ❖ ❖ ❖ ❖ ❖ ❖ ❖ ❖ ❖

As late as 1920, Rhode Island Greening was in the top six apples grown in the U.S., and in 1900 it probably slightly surpassed the Baldwin in numbers of trees grown in parts of New York state. Being less hardy than the Baldwin, Rhode Island Greening needs the more favorable locations in that state. By 1990 it was still in the top 15 commercial apples grown in the U.S., with production just over three and one- half million 42-pound boxes per year. It hovers in the top 15 list, from 9th to 14th, vying from year to year with Gravenstein, Northern Spy, Newtown and Idared. Much of its production today is processed for pie or sauce but, because of a peculiar and attractive flavor, Rhode Island Greening is above average for fresh dessert and it is in good demand in those localities that produce it.

According to Beach and older writers, this cultivar no doubt originated in the State of Rhode Island, probably in the vicinity of Newport, near a place known as Green's End. In 1898 in the town of Foster, on the farm of Thomas Drowne, stood a Rhode Island Greening tree that was supposed to be 200 years old, and some believed that perhaps it was the original tree. In 1898, scions (grafting twigs) from this tree, as well as scions from a sucker growing up from the roots, were brought to the N.Y. Agricultural Experiment Station in Geneva, New York. If grafts from both the root and the tree were to fruit the same, then it would be proof that it was the mother tree, planted from seed. They did not fruit the same. The grafting wood from the tree proved to be the regular Rhode Island Greening, but the grafts from the roots were not. Therefore, this tree on the farm of Thomas Drowne was not the original tree, and those who said that the mother tree would have been a good deal older than this tree were probably correct.

Professor S. A. Beach was the chief horticulturist at the Geneva Station at the time of the above-noted test, and he wrote about it in Volume I of *The Apples of New York*. One has to assume he was very interested in testing the grafting wood from that old tree, and he gave a long, five-page account of this apple, with many older references. Due primarily to considerable research by Beach, Hovey and earlier pomologists, we have information on Rhode Island Greening that is not really speculation.

The age of this cultivar has been pushed back to around 1650, regardless of whether the original was the tree at Green's End Inn or a tree near the one at Green's End Inn.

In our current over-populated, over-industrialized big city environments, it can be difficult to understand the environment around the original Rhode Island Greening trees. Most of the population of Rhode Island then lived in villages and, to fix the time period in your mind, it was quite possible that Miles Standish of the New World Colony was still alive, as he died in 1656, having come to America on the Mayflower in 1620. This period was almost one and one-half centuries before our Revolutionary War, and to someone of that time the Appalachian Mountains were part of a dangerous wilderness just to the west. Very few of these people were skilled in grafting, so thousands of apple seeds were planted each year, many having been brought from Europe. Since cultivars do not breed true from seed, it was seldom that one obtained an apple anywhere near the quality of Rhode Island Greening.

Since communication was by word of mouth, and transportation was by horse or horse and buggy, it is not unusual that the Rhode Island Greening cultivar was not grown very widely for almost 100 years. It was introduced into the old Plymouth Colony from Newport around 1760 and was taken to the new Ohio country in 1796 by General Putnam.

In 1922, Hedrick stated much the same description as Beach; i.e., that Rhode Island Greening was an excellent dessert apple, and was unsurpassed for culinary use (pie, sauce, etc.). Its fine qualities have kept it a popular apple for centuries.

The tree grows slowly but, like many other triploids, eventually it becomes very large. It has one of the longest life spans of all apple trees. There are many trees of this cultivar in the eastern U.S. that have reached 150 years of age on seedling root (standard rootstock).

The western U.S. has never grown many, and I have only seen Rhode Island Greening trees twice in Washington or Oregon. Six miles from my home is the oldest incorporated city west of the Rocky Mountains, Oregon City, Oregon, located on the Willamette River near a waterfall. An old estate of five acres is located on the west side of Oregon City, and in 1980 its elderly owner asked me to identify an apple she said was her favorite all-purpose apple. It was not difficult to identify it as a Rhode Island Greening. I mention this because the owner said the Rhode Island Greening tree was supposed to be in excess of 60 years old when she bought the estate in 1941. This means that this living tree is now well over 100 years old, and it was planted when areas of Oregon were very much in pioneer days. Since the owner's death, the estate has been designated an historic site.

For those who scoff at this old apple, ask yourself, will Red Delicious, now 100 years old, still be in the top 15 commercial cultivars in the year 2220, 230 years from now, as is the Rhode Island Greening? Maybe.

Rhode Island Greening may never be a favorite here in the west for the home orchard grower or the commercial grower. Here, our old historic cultivars are Newtown, Spitzenberg, Northern Spy, Jonathan and Gravenstein, all of them regularly grown to perfection in the Willamette Valley and Hood River area, at least by homeowners.

Rhode Island Greening is too sensitive to cold weather to grow well in upper McIntosh country, and it does not do well in the south, where it becomes a fall apple and loses quality.

It appears, however, that Rhode Island Greening will continue to be a favorite of the eastern U.S. growers, partly because it originated there and its history is there. In 1990, Rhode Island Greening is still the most planted or second most planted apple in certain parts of New York state, with much of the crop used for processing, fulfilling a valid need.

Beach said it was unsurpassed as a cooking apple and, for comparison, the U.S. was growing thousands of cultivars in 1903. I have grown every apple in this book with a colored picture (except Wickson), and only Spitzenberg, Newtown, Gravenstein, Elstar, Northern Spy, and Calville Blanc d'Hiver and perhaps Bramley, would be in the same league for quality as a cooking apple; all except Elstar have difficult-to-grow tree characteristics. When ripe, Rhode Island Greening is also a fine dessert apple with a very unique flavor.

Many of us who test apples and the trees often scoff at something old, and for no other reason except it is old. I hope I am not that kind of tester.

Arlet (Swiss Gourmet™)

Ashmead

Blushing Golden™

Braeburn

Gala

Ginger Gold™

Golden Delicious

Granny Smith

Gravenstein

Grimes Golden

Haralred

Hudson's Golden Gem

Idared

Jonagold

Jonathan

Kandil Sinap

Karmijn de Sonnaville

Keepsake

Kidd's Orange Red

Liberty

McIntosh

Melrose

Mutsu

N.Y. #429

Four-year old apple trees on size controlling rootstock, trained to a 4-wire trellis.

Typical Codling Moth damage, with worm (larvae) exposed.

Apple Scab shown on fruit and apple leaves.

M-9 rootstock chewed off by voles (large short-tailed mice).

Flathead Apple Tree Borer damage in graft union of young apple tree (Mark Rootstock.)

Trunk of the Hawkeye Delicious tree near Peru, Iowa, that remains from the two trunks that grew back from the original tree in the 1940's, after the original trunk was killed. Picture by Bob Denney of NATCO, June 1995. (See description of Delicious.)

Well-pruned "Central Leader" tree in author's test orchard. (Newtown cultivar.)

15-year-old , 4-wire trellis; often, this is called a tree wall.

4-wire trellised row in the author's yard. (Netting was to protect from bird damage.) Three trees have M.7 roots, one has M.9.

Fake owl guarding near-ripe Elstar from birds. It worked!

Wild honeybee collecting pollen from an apple blossom in author's yard.

Roxbury Russet

*

Tree:

In Ohio, the **synonym** before 1800 was Putnam Russet. **Originated** at Roxbury, Massachusetts, just before 1649. **Parentage** is unknown. Chromosome count is 2 x 17 = 34, a **diploid**. It has some self-fertility and is an excellent pollinizer in the mid- to late-bloom.

Tree is of medium vigor and **size**. Bearing habit is usually **annual** if thinned, sometimes slightly biennial. **Winter hardiness** is adequate in USDA zones 5 and 6 with trial in protected lower zone 4. Farther south the tree grows well but, according to Beach, apple quality is lessened. Here in the Willamette Valley's zone 7 (maritime) climate, it grows to perfection. Melrose blooms at the same time and they cross-pollinize very well. It has **disease resistance** to scab and fireblight, and is quite resistant to powdery mildew.

Sources: small nurseries specializing in antique apples, some listed in the back of this book. Roxbury Russet is the oldest known American apple of any note. Since trees (or grafting wood) were taken from the Roxbury area of Massachusetts to Connecticut in 1649, then the original tree probably fruited between 1625 and 1640. The seed for the original tree would almost surely have had to come from England, as that early on the eastern seaboard there were not many (or none) cultivated apples to supply seed.

Fruit:

Size is medium to medium large. **Shape** is variable on the same tree, with about 20 to 50 percent elliptical in shape, a rarity. (See the drawing on apple shape.) **Skin color** is always two and sometimes three shades, varying between green, plus green-gray russet. It is a splotchy mixed color apple that I find very different and even attractive.

Stem length is medium to short, thick, and often covered with brownish-orange pubescent. **Flavor** is different and sprightly, slightly tart; however, the apple has a high sugar content and apple juice made from this one cultivar is quite sweet. **Flesh texture** is somewhat coarse, greenish to yellow in color.

Maturity is from October 10 to 30 across northern zones 5 and 6 of North America. **Storage life** is quite long even under garage conditions, with four to five months in common storage, 30° to 32°F. No **physiological disorders** noticed.

Quality rating: **Good to Very good.**

❖ ❖ ❖ ❖ ❖ ❖ ❖ ❖ ❖ ❖ ❖ ❖ ❖ ❖ ❖ ❖ ❖

Because of its qualities, Roxbury was taken to the New West—Ohio in 1797—where it was called the Putnam Russet.

Those who have not grown Roxbury Russet and Golden Russet may think they are very similar. They are not; there are discernible differences. Roxbury is larger and many of the apples are elliptical. Golden is round and usually 20 percent smaller. Roxbury varies a great deal in color on the same tree, while Golden is more uniform in color and size.

Golden Russet has much of its fruit growing on branch ends (a tip bearer). Roxbury blossoms are throughout the tree. Golden is richer and tastes more sweet and, apparently, has less acid as Roxbury has much sugar but it is not as apparent.

Before the use of refrigeration, Roxbury was shipped extensively to the south and west and all the way to the West Indies, off Florida. It has outstanding keeping qualities as many russets do.

Its flavor is very different, sort of piney, but I like it and would eat it more for dessert except for comparing it to many of the world's best dessert apples I grow. Now, I keep it primarily for an apple juice mix. If Roxbury, Newtown, and one other cultivar, perhaps Spitzenberg, are mixed, the cider is fit for royalty.

I also like to leave about 30 fruit on the tree to attract some of my favorite feathered friends, such as the varied thrush (Oregon robin), flickers, towhees, robins, etc. The carbohydrates are helpful to their survival in bad weather.

If the reader wants a long keeping and fine cider apple that is at least 350 years old, plant a Roxbury Russet. Don't forget its resistance to scab, fireblight, and mildew.

Spigold

✳

Tree:

Origin was at the State Agricultural Experiment Station in Geneva, N.Y. It was selected in 1953 and introduced in 1963. **Parentage** was Red Spy x Golden Delicious. Chromosomes are 3 x 17 = 51, a **triploid**.

A vigorous grower and a rather large tree except when grown on the smaller rootstocks. **Biennial** in bearing habit. **Winter hardiness** is unknown, but I suggest growing in USDA zones 5 to 8. **Disease susceptibility:** apple scab fungus and certain cankers. May be susceptible to fireblight. **Blossom season** is late mid-bloom to well into the late-bloom. Poor pollinizer because of chromosome count.

Sources: widely available from many nurseries. A very erratic tree for commercial production and seldom grown in any but small orchards, home or commercial. Slow to come into bearing.

Fruit:

Size is large to very large, often one pound in weight. **Shape** is an irregular round-oblong and is usually distinctly ribbed as is one of its parents, the Red Spy. **Stem** is fairly long and stout. Usually not as long or slender as Jonagold. Does not seem to **russet**. **Storage life** is short, not over two months. Matures in mid- to late-October here in USDA zone 7 of the Willamette Valley.

Flesh is ivory-white, with crisp fine texture and of above-average flavor. **Color** is an attractive red over a yellow-green background. Excess nitrogen coupled with warm pre-harvest nights can give rather colorless fruit with ordinary **flavor**. A **physiological disorder** is bitterpit, especially on trees under six years or so.

Quality rating: **Very good to Best.**

◇ ◇ ◇ ◇ ◇ ◇ ◇ ◇ ◇ ◇ ◇ ◇ ◇ ◇ ◇ ◇

As grown here, it has been difficult for me to distinguish Spigold from Jonagold at fruit shows, unless one has a number of specimens of both. Its shorter, stout stem usually separates it from Jonagold. Occasionally a very large specimen also gives it away. It even tastes like Jonagold, although at its best, I think it is better.

Years ago, before knowing much about either, it seemed Jonagold got all the publicity. Now it is apparent Jonagold's better tree characteristics are the reason. Spigold is an irregular bearer like one of its parents (Red Spy), and it is more difficult to contain than the Jonagold tree. It simply does not produce those near annual crops like Jonagold. Both cultivars are triploids with poor pollen. As with Red Spy, the large apples on young trees are quite susceptible to bitterpit.

Its very large size seems to turn off the packer-sales organizations. It appears a uniform size and standardization are more important to them than flavor, and perhaps rightly so.

Spigold is an all-purpose back-yard apple for the connoisseur and probably will never make the stringent qualifications of storage and shipping. Direct marketers or specialty shops should find its large size and great taste quite profitable. It is in this book because it is the best tasting large apple I have grown. It generally does not keep well past one or two months in regular cold storage of 30° to 32° F.

Spitzenberg

❋

Tree:

Synonyms: Esopus Spitzenberg, Spitzenburg. A **diploid** of 34 chromosomes of unknown **parentage.** I suggest cooler areas of USDA zones 5 through 7 and warmer parts of zone 4. **Ripens** in mid- to late-October in western Oregon.

Tree is hardy and of modest **size.** Without protection, it is very **susceptible** to scab, mildew, fireblight and canker. **Biennial** if not thinned.

Sources: many specialty nurseries such as Bear Creek, Fruit Testing Association Nursery, Inc., Northwoods, Living Tree Center, Raintree, and others.

Fruit:

Shape is variable, usually oblong-conic. **Size** is below medium to medium-large. **Color** is red with yellow dots (with low nitrogen). **Stem** is medium to short and often knobbed. The dots are much more numerous at the calyx end of the apple than on the top half. This is an aid to identification. **Stores** for three to five months at 32ºF.

Quality rating: **Very good to Best.**

◆ ◆ ◆ ◆ ◆ ◆ ◆ ◆ ◆ ◆ ◆ ◆ ◆ ◆ ◆ ◆

In Beach's *The Apples of New York,* he says that when well-colored, Spitzenberg is unexcelled in flavor and quality. That is true even in 1995. He also states it was well known in New York and adjoining states more than 100 years before (from the early 1700s).

I was always disappointed that Beach said he could not find authentic accounts of the time of origin. It appears Spitzenberg was discovered in the early 1700s at Esopus, on the Hudson River 80 miles north of New York, presumably by an early Dutch settler named Spitzenberg. It is apparently difficult to grow Spitzenberg, at least to perfection, probably because of disease susceptibility, in a humid summer climate such as that of Virginia, although Thomas Jefferson and others planted a great many of them.

The Willamette Valley of Oregon, where I live, is one of the very best places to grow and ripen Spitzenberg. I have learned to control fungal infections by spraying buds with anti-fungal sprays just as buds start to swell, and do this a number of times until the rains stop in May or June. The picture shows their beautiful color about October 20 as grown in my orchard on M.9, which would be my first choice of rootstock. Apple anthracnose seems to be well controlled by painting a strong solution of copper sulfate when it appears on branches and trunk. Scab and mildew can also be controlled, and we do not have to worry about fireblight in our valley.

Perhaps we could revive this marvelous tasting apple, even though it does seem site selective. Areas with warm (not hot) days, dry summers and cool September and/or October nights for good color, plus a long frost-free fall appear to be the best sites, but such areas are limited.

Spitzenberg is still a terrific all-purpose apple for fresh dessert, baking, drying, apple sauce and pie. It will probably always be one of my top four favorites. A few direct-market growers in our valley grow and sell Spitzenberg for good prices, but other areas seldom grow it commercially anymore, probably because of disease, and three generations have been brought up on lesser fare.

Spitzenberg is an antique apple that was a prize in the 18th, 19th and 20th centuries, and now most certainly for the 21st century, at least for connoisseurs.

Stayman

❄

Tree:

Synonym: Stayman Winesap. Chromosomes are 3 x 17 = 51, a **triploid.** Vigor and tree **size** are moderate. **Annual** in bearing habit and tends to overset; some thinning should be done. **Winter hardiness** is medium and should be grown in the mid-latitudes of USDA zones 6 through 8 for best quality.

Quite **disease free,** but slightly susceptible to apple scab fungus; fairly resistant to fireblight. **Blossom** season is the mid- to late-bloom.

Sources: most commercial nurseries.

Fruit:

Size is medium to medium-large. **Shape** is conic-round with much uniformity in shape. **Skin color** is light red, with darker broken stripes over ground color of yellow-green. **Stem** length is short to medium. **Flavor** is sweet, sprightly and less sharp than its seed parent Winesap. **Flesh texture** is firm, crisp, moderately fine grained with yellow to yellow-green color. It **russets** frequently on the stem end of the fruit.

Pick in mid- to late-October in zones 6 through 8. **Stores** three to five months in common storage at 30° to 32°F. **Physiological disorder:** some strains crack on the apple shoulder in about 10 to 15 percent of the apples. Poor management, soil, pH, and climate contribute to cracking.

Quality rating: **Very good.**

◊ ✧ ◊ ✧ ◊ ✧ ◊ ✧ ◊ ✧ ◊ ✧ ◊ ✧ ◊

Dr. J. Stayman of Leavenworth, Kansas, planted a number of Winesap seeds (open pollinated), apparently in 1866, and when they fruited about 1875, one of them proved to be an outstanding new apple. It was soon to be named Stayman.

Even in 1903 when Professor Beach and his colleagues had written *The Apples of New York*, it was not horticultural practice to attach the name of one parent to a seedling. Dr. Stayman, and perhaps the horticulturist Downing, probably were the first to call it the Stayman Winesap. For some unknown reason, Beach copied their terminology. Stayman is really its correct horticultural name.

Stayman is larger, lighter colored, sweeter, and more disease resistant than its parent. It is not a better keeper than Winesap, but then not very many apples are.

Like Jonagold, Stayman is another triploid tree of moderate vigor and size, and exhibits little of the large, heavy, stiff growth some triploids such as Gravenstein, Holstein and Baldwin show. I grow all of these so can easily compare these cultivars in the same environment. My triploids are grown primarily on M.9 or M.26 rootstock and they are all above average in quality. Pollen sources are nearby and plentiful.

Stayman does well in various climates and soil conditions in the mid-latitudes, unlike its parent Winesap. Its main production area is similar to that of York, which extends from New Jersey, the Virginias, Delaware, Maryland and westward through Pennsylvania. Because, like the York, it is used a great deal for processing, growing it in the same areas as York makes it easier to process.

I know of only one grower who has any quantity of Stayman trees here in the Willamette Valley, and because of the premium prices he receives (selling them as fresh dessert apples), he calls his Stayman trees "money trees."

If, after 115 years, the west (meaning the portion of the U.S. west of the Mississippi) has not adopted this fine old apple, then it probably never will. Our lack of plants that process apple slices and applesauce may be part of the reason. At least some of us have a tree in our back yard so we can savor this fine fruit.

Sturmer

✳

Tree:

Synonym: Sturmer Pippin. **Origin** was in England about 1827 and it was introduced in+ 1843 or later. **Parents** are thought to be Ribston and Non Pareil, two above-average apples. Chromosome count is 2 x 17 = 34, a viable pollen **diploid**. Medium **size** tree. **Annual bearer. Hardiness** is unknown in the U.S. because of its rarity.

Disease: somewhat susceptible to apple scab fungus and canker diseases in damp, mild winter climates. **Bloom** time is mid- to late-season. I suggest growing in USDA zones 6 through 9, with cooler ares of zone 6 perhaps too short a season.

Source: Rocky Meadow Nursery.

Fruit:

Size is medium to large. **Shape** is usually oblate (flattened) to round-conic. **Skin color** is green with in some seasons a brownish-red flush, frequently **russeted** on both ends of the apple. **Flavor** is very tart, almost sour, until ripe, when it has a rich, high flavor. For best flavor, never pick it early. **Stem** is short, thick, and to the top of the stem cavity. **Flesh** is white.

Matures in early November in zone 7 of the Willamette Valley, Oregon, sometimes earlier. Fruit hangs well until mature. I suggest **storage** at 30º to 32ºF for one month before eating. **Physiological disorders:** sometimes gets bitterpit, a calcium shortage in the apple, especially on young trees. An all-purpose apple.

Quality rating: **Good.**

❖ ❖ ❖ ❖ ❖ ❖ ❖ ❖ ❖ ❖ ❖ ❖ ❖ ❖ ❖

Sturmer was raised by a nurseryman, Mr. Dillistone, at Sturmer in the Suffolk area of England about 1827.

Because Sturmer requires such a long, warm and late season to ripen properly, one could probably not find a commercial grower in England today. The climate is too cool in its birthplace, or it would certainly vie with Bramley as a favorite cooking apple. I like it better than Bramley because it seems to have more sugar and is a fine tasting apple when picked in this valley in early November. It is a very good all-purpose apple with rather fine-grained white flesh.

Old writers in England extolled its flavor, but it ripened there only in the occasional long hot summer. Fortunately, the pioneer fruit growers of Australia proper, Tasmania, and New Zealand took it "Down Under," where it is a fine success, and quantities are shipped to the British Isles from late May to August. Their reverse seasons put a fresh apple on England's grocery shelves just ahead of local Cox Orange and Bramley.

Some years ago an acquaintance returned from a summer trip to England and said he had just tasted a very good apple that was new to him. Of course, I had read of it for many years but I had not done anything about it. Now, I have fruited it for five years on MARK rootstock and am sufficiently impressed to include it, especially for the longer season areas. The southern half of the U.S. would be an appropriate area in which to try growing Sturmer.

American commercial growers probably will not grow it as "not attractive enough." This is for the connoisseur who lives in USDA zones 6 through 9. We need more apples of quality for the southern part of the U.S. that ripen after the weather starts cooling off, and this is one of them.

Except for my friend's Sturmers and mine, I know of no one growing it in this area, so it certainly is not very well known. It is somewhat similar to Karmine in flavor, and should be grown one or two USDA zones farther south than the northern limit of Karmine. If you can ripen Arkansas Black, Granny Smith or Braeburn, it would be worth a try.

Sweet Sixteen

❊

Tree:

Origin was a 1937 cross of Minnesota #477 (open pollinated Malinda) x Northern Spy by W. H. Alderman at the Excelsior Station of the University of Minnesota. A **diploid**, 2 x 17 = 34 chromosomes. **Size** of tree is moderate and of moderate vigor. **Annual bearer** if thinned. **Winter hardiness** through USDA zone 4. I suggest zones 4 through 6. **Disease:** resistant to apple scab fungus, fireblight, and cedar apple rust in Minnesota. **Blossom season** is in the mid-bloom.

Sources: Bailey Nursery and Rocky Meadow Nursery.

Fruit:

Size is medium to medium-large. **Shape** is round-conic. **Skin color** is rose-red overlaid with red stripes over a green to yellow ground color. **Stem** length is medium to medium-long. **Flavor** is very unique and a sweet-tart cocktail of flavors. **Texture** is fine and crisp; **flesh** is yellow-white when ripe. **Russets** around the stem more often than not. **Matures** in southern Minnesota between September 19 and 27 and about the same time in western Oregon.

Stores well until mid-December at 30° to 32°F. No **physiological faults** known. A quality north-country apple.

Quality rating: **Very good.**

✦ ✦ ✦ ✦ ✦ ✦ ✦ ✦ ✦ ✦ ✦ ✦ ✦ ✦ ✦

In protected areas or near large bodies of water on cold-hardy rootstock, this new cultivar could prosper in some of zone 3. It is suggested by the University of Minnesota for trial in the northern half of Minnesota, which is in USDA zone 3.

W. H. Alderman at the University of Minnesota Horticultural Research Center in Excelsior, Minnesota, was an outstanding apple breeder in the 1930s and 1940s. Minnesota #477 originated from an open pollinated seed of Malinda; Sweet Sixteen was selected in 1950 and released in 1978. Malinda was used in breeding by the University of Minnesota because of its cold-hardiness. It was an obscure cultivar that was brought from Vermont to Minnesota about 1860. Eventually, from Malinda came the hardy Haralson (and Haralred), which is now the most grown apple in Minnesota and the first Malinda offspring of note. So Sweet Sixteen is theoretically one-quarter Malinda, inheriting much of its cold hardiness from this grandparent. Certainly it also inherited some cold hardiness from Northern Spy. Sweet Sixteen was tested for 28 years across Minnesota's nearly 400-mile length before release.

The work of W. H. Alderman is especially notable because some excellent cold-hardy and good quality apples have resulted from his efforts and those of his colleagues. Perhaps no other apple breeding program except the Morden Station in Canada has had as much success with cold-hardy cultivars. Sweet Sixteen and Honeycrisp stand somewhat alone on the quality scale among the north country offerings. They have above average quality.

Growers in zone 3 need a rootstock that provides more vigor and cold hardiness, such as Beautiful Arcade. In some Canadian tests, Beautiful Arcade has been shown to have exceptional cold hardiness. It is a very old (before 1800) Russian apple. Borowinka (Duchess of Oldenburg) and Columbia Crab rootstock should also survive zone 3. (See "Cold-Hardy Rootstock.")

My personal opinion is that this apple is of higher quality, with much better tree and fruit characteristics, than McIntosh. After tasting Sweet Sixteen apples that were grown in Michigan, Minnesota and Oregon, I have to place this cultivar in the "sleeper" category—although it already has a following among connoisseurs from the Midwest to the East Coast.

The southern growing limit for Sweet Sixteen is not known to me. The fruit ripens well in zones 4 through 6 and is still relatively unknown in the western part of the U.S. My tree is nine years old as of this writing.

Tydeman's Late Orange

✳

Tree:

Synonym: Tydeman's Late Cox. Introduced in 1949, about 20 years after crossed. A **diploid** of 34 chromosomes. A vigorous grower, but **modest-sized** to almost small tree with long, weeping, slender branches. **Winter hardiness:** I suggest USDA zones 5 to 8, (with trial in zone 9, especially above 2,000 feet of altitude.) **Disease** resistant to mildew and quite resistant to apple scab fungus. **Blossoms:** mid to late bloom.

Sources: Southmeadow Fruit Gardens, Northwoods, Bear Creek.

Fruit:

Size: medium-small if not thinned early. **Shape** is conic to round-conic. **Color** is a greenish-yellow ground flushed with orange-red stripes. Most years it has fine russeting on the upper half. **Stem** length is short to medium and usually stout, but is variable and can be thin. **Flesh texture** is juicy, fairly firm and cream to yellow-colored, depending on ripeness. **Matures** here in mid- to late-October but will hang well on the tree much longer.

Storage life at 32ºF is three to four months if not picked too late. Better **keeper** than Cox Orange. Quite **disease resistant** and no **physiological disorders** in our climate.

Quality rating: **Very good to Best.**

◇ ◇ ◇ ◇ ◇ ◇ ◇ ◇ ◇ ◇ ◇ ◇ ◇ ◇ ◇ ◇ ◇

H. M. Tydeman was a distinguished researcher at East Malling Research Station in England and around 1930 made the cross resulting in this apple.

Tydeman's Late Orange should not be confused with another cross by this same researcher, Tydeman's Early (Tydeman's Red), a McIntosh cross that is certainly not in the same class as the Late Orange.

My test orchard has this cultivar on M.7A rootstock and Budagovski 491 (B.491). It heavily oversets, so after 15 years of experience with it I now understand how much thinning it needs in the up year (at least 80 percent thinning) to keep it from strong biennial bearing. We all liked the apple so much, sheer greediness formerly kept me from enough thinning; a short-sighted affliction I do not recommend.

Lineage listed below is interesting, because Cox Orange was one grandparent and also one parent.

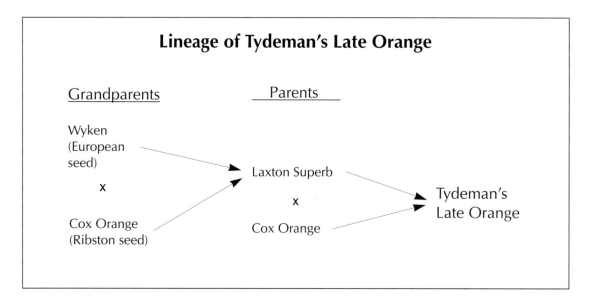

Lineage of Tydeman's Late Orange

Grandparents　　　　　Parents

Wyken
(European
seed)

x

Cox Orange
(Ribston seed)

Laxton Superb

x

Cox Orange

Tydeman's
Late Orange

Despite a strong biennial bearing trend, this is a much easier apple to grow than Cox Orange. I have seldom seen the cracking one encounters the first few years on Cox. In climates that are hot during pre-harvest, Cox Orange frequently gives poor fruit color, too early ripening and mediocre flavor, whereas Tydeman's Late Orange seems to slide through the warmer July and August weather, ripening after the weather has cooled off.

If you live below a line extending from Chicago to Boston, then one may want to try Tydeman's Late Orange instead of Cox Orange, as chances of getting high-quality apples are much better. I see little difference in their quality in my climate, and both do well here but Cox is a skimpy, shy bearer.

Tydeman's Late Orange is perhaps the most overlooked English apple in America. Grown commercially in a limited way in England, it will probably never be grown commercially in the U.S., except by direct marketers. Its rich flavor makes it a connoisseur apple of the first order.

Wickson Crab

✳

Tree:

Origin by Albert Etter, private breeder, probably in the late 1920s or early 1930s, in Humboldt County, California. **Parentage:** two crabs, the Spitzenberg crab x Newtown crab. Chromosomes are 2 x 17 = 34, a **diploid**. Somewhat **susceptible** to apple scab fungus.

A vigorous and fairly **large** tree on seedling rootstock; suggest using size- controlling rootstock. Heavy **annual bearer**. Needs thinning for the larger fruit. Bears on one-year branches (the last summer's growth.) **Winter hardiness** is quite adequate and should perform well in USDA zones 4 through 8. **Blossom** season is early- to mid-bloom.

Sources known to me are Greenmantle Nursery and Southmeadow Fruit Gardens.

Fruit:

Size when well thinned is just smaller than Whitney Crab. If not thinned, they could be only 1" to ½" in **size**. **Shape** is oblong to conic, and quite large for a crab apple. **Color** is bright red over yellow ground color. **Stem** length is medium to short. **Flavor** is an excellent sweet-tart, with high sugar content. Texture is crisp and juicy, and fruit exhibits a small core. It does not seem to russet.

Matures in northern California and western Oregon from October 10 to 20. Tends to bruise easily and such fruit would have a short **storage life**. No **physiological disorders** known.

Quality rating: **Very good to Best** among edible crabs.

❖ ❖ ❖ ❖ ❖ ❖ ❖ ❖ ❖ ❖ ❖ ❖ ❖ ❖ ❖

This is one of the famous "Etter apples." Albert Etter (1872-1950) stands alone in using wide, imaginative and exciting apple crosses, frequently with a crab as one parent. In the early 1900s, he was a strawberry breeder of note among the hills of Humboldt County in northwest California. Dr. George Darrow, a USDA berry plant breeder, gave Etter a prominent place in the world of strawberry breeding in the 1937 USDA yearbook, as well he should. Around 1910 to 1918 Etter had started crossing genetically diverse apple parents, and he gradually abandoned his strawberry work. In 1929 he turned all of his strawberry breeding material over to the University of California.

A more or less self-taught, later-day "Luther Burbank" type, Etter never got the credit he deserved from the establishment for his apple breeding. After all, who else of that time period has at least a dozen above-average quality apples credited to their work, with another eight or so pink-fleshed varieties not yet even named.

Ram Fishman, owner of Greenmantle Nursery at Garberville, California, and former president of North American Fruit Explorers, lives near the old Etter homestead. He says in this homestead and others nearby, he is trying to rescue any number of unnamed and potentially above-average apples: crabs, pink-fleshed types and regular-sized apples. We may be in the 21st Century before we really know the extent of Etter's apple contributions. I have long maintained that horticulturists should have some special genetic dispensation given them for a life of at least 150 years, so they can see the fruits of their labors better rewarded. They could also then carry on with their own research. Albert Etter died almost in poverty, and I think he would have agreed with me.

Wickson Crab was named after E. J. Wickson (1848-1923), whose book *California Fruits*, published in 1889, remains a bible for historical material. E. J. Wickson was Etter's friend and perhaps mentor, so it is fitting it should be named after him (as was a very good plum developed by Luther Burbank.) Before his death, Wickson had become dean of the College of Agriculture at the University of California.

Why a crab apple?

Many people think of crabs as just showy ornamentals, or something such as Dolgo, from which people made wonderful-tasting jelly years ago. Most crabs are excellent pollinizers of other apples and since many crabs have *Malus baccata* genes in them, some horticulturists go a step further and recognize them as cold-hardy rootstock (Columbia and Dolgo).

There is another, but small, group of crab apples that are really all-purpose. They also have attractive blossoms, are quite cold-hardy and usually are genetically dwarfing. Fruits from this limited group are from one and one-half to two inches in diameter and are superior for fresh eating and in pickling. They are especially desirable in fresh apple cider. I think this small group of crab apples is much overlooked because they often do not keep until the October or November fruit shows and have little exposure. Also, the average family now does less fruit cooking or jelly making, and they also make less apple juice cider than 75 years ago.

In any event these very useful apples, big for crabs, should be better known. Besides Wickson, which I think the best of the fine crabs, somewhat in order of my preference are: Young America, Centennial, Whitney and Martha. Of this group, I am just featuring Wickson.

Mr. Fishman of Greenmantle Nursery courteously supplied me with a copy of the Wickson Crab patent papers. They were applied for in June 1944 and patented in March of 1947 (#724), now expired of course.

The only other description I have seen of Wickson Crab said its parents were Yellow Newtown and Spitzenberg apples. This is very much in error, and is not quite what the patent paper says. Under origin, Etter's application says: "It is a cross of Newtown *crab* and Spitzenberg *crab*." (Emphasis mine.) The problem here is trying to find information about these parental crabs. Ram Fishman says one of these parents is mentioned in a book in 1900; I speculate that the other may have been a crab produced by Etter, as he made a myriad of crosses. In any event, Wickson is not the offspring of two of our best old apple cultivars, Newtown and Spitzenberg. Obviously these names should not have been used for the crab parents; other names would have been much less confusing.

Because of its high sugar content, reputed to be 25 percent, Etter thought it would be useful for apple champagne. He proved its worth as sweet apple cider by bringing a few gallons to local fall neighborhood festivities, where people enjoyed it and quickly drank it all up. It is one of the few apples that needs no mixing with other cultivars for superior apple juice.

In the "All About Fruit" Shows in 1990 to 1993 in western Oregon, considerable quantities of Wickson Crab were on display, so we have finally discovered this sweetmeat here in Oregon.

Etter said of his new variety in the patent application: "This diminutive apple surpasses most crab apples in color, form and flavor and masses of red fruit almost conceal the foliage in fall." I agree that Wickson is an excellent crab apple. I hope the reader would have a garden space where he would like to try a crab apple, perhaps Wickson.

Winesap

✳

Tree:

Synonyms: Holland Winter or Texas Red. **Origin:** see text. Chromosomes are 2 x 17 = 34, a **diploid**; however, pollen is quite infertile so it is a poor pollinizer. A rather slow grower of moderate **size** but quite vigorous. Leaves are narrow and small to medium when compared to a majority of apple trees. One of the few really **annual** bearers without very early thinning, but it needs thinning for the apples to size.

Sources: Most commercial nurseries.

Fruit:

Usually uniform in **size** and, when grown in a long season climate, fruit is medium to sometimes medium large; however, in much of upper New England and in USDA zones 5 or 6, fruit tends to be small. **Stem** is medium to short and rather slender. **Skin** is medium thickness but tough. **Color** is glossy red with stripes of purple red over yellow (or green if immature) ground color. The prevailing effect is usually bright, deep red. **Flesh** is yellowish white, often with red veins and a tart flavor. **Texture** is firm, crisp and somewhat coarse. **Shape** is conic.

Quality rating: a tart **Good to Very good.**

❖ ❖ ❖ ❖ ❖ ❖ ❖ ❖ ❖ ❖ ❖ ❖ ❖ ❖ ❖ ❖

This is one of the oldest and at one time one of the most popular apples grown in North America. It is not known for certain if the cultivar originated in this country. It is possible that grafting wood or seeds were brought over by ship from northern Europe or England to the New Jersey/Maryland area before the Revolutionary War. On the other hand, millions of apple seeds were planted in North American from 1650 to about 1880. Some of the seeds were brought from Europe, so it seems more likely that the Winesap cultivar is one of those seedlings, probably of American origin.

In 1817 Cox, the first American to write a book about apples, speaks of it as being the most favored cider fruit in West Jersey. Winesap then is probably from the colonial era and is perhaps 210 to 230 years old.

According to Bailey's inventory of 1892, 73 nurseries offered grafted Winesap apple trees for sale. This compares favorably with other apples, as 64 nurseries offered Baldwin, 58 offered Northern Spy, and 48 offered Rhode Island Greening. There is evidence that Winesap was not as widely planted commercially as were the latter three, so many homesteads and farmers must have planted small numbers of Winesap to support such wide availability. In other words, homeowners liked it and some do yet.

Winesap has two great qualities: it gives an annual and abundant crop. In addition to producing a superior, tart fruit, it is a lovely ornamental with showy pink flowers.

Because this apple does not grow well on seedling rootstock in poorly drained or heavy clay soil, it will do no better in such conditions on Malling or Malling Merton rootstock. One exception might be M.13, which is semi-standard. M.13 rootstock is rather scarce, being available from two Oregon rootstock nurseries. M.13 with its fibrous, shallow roots survives in heavy, wet soil better than most clonal rootstocks and produces trees slightly larger than those on M.7A rootstock, but it is drought susceptible. Growing conditions are best for Winesap in light soils in USDA zones 6 and 7.

In 1988, half the national crop of Winesap came from zone 6 in eastern Washington, where about 1.4 million 42-pound boxes were produced. Moosebar, which is quite red, is the predominant strain used in this region.

The commercial popularity of Winesap has waned primarily because of its fine offspring, the Stayman, a larger, less tart-tasting apple, and Stayman's production is almost double that of its parent. The huge production of Red Delicious has also affected Winesap's decline, as well as that of other cultivars.

If Winesap is cooked with the skin on and pressed through a colander, the apple sauce develops a natural pink-red color. Like Wealthy, Winesap often has red streaks through its flesh. For 200 years it has proven outstanding in sweet cider mixes. Winesap is often grown in eastern Washington specifically for apple juice; however, due to low prices offered the grower, production is declining commercially.

Also like the Wealthy, Winesap's foliage is somewhat sparse. On fertile, light loam soil, the fruit is quite uniform and of medium size. According to Beach, when grown farther north, as in New York, fruit size is usually small.

Because of some excellent newer cultivars such as Gala, Jonagold, Braeburn, Elstar, and Fuji, it is probable that Winesap will continue to decline, just as have Spitzenberg and Baldwin.

Winesap has two notable seedling descendants, Stayman and Arkansas Black. Both were originally rated "Good to very good" by horticulturists Beach and Hedrick, and I would so rate them today as middle latitude (long season) cultivars.

Winesap has a diploid chromosome number of 34; however, because of very defective pollen, it acts like a triploid of 51 chromosomes and should not be relied on as a pollinizer. It has excellent keeping qualities under garage storage conditions, and it is seldom put under controlled atmosphere (C.A.) storage because it keeps so well in regular storage of about 30° to 32°F.

A reliable and desirable apple for the domestic orchard when grown in zones 6, 7, or 8, and especially desirable for apple juice mixes.

York

*

Tree:

Synonyms: York Imperial, and Johnson's Fine Winter before 1850. **Origin** was on the Johnson farm near the borough of York, Pennsylvania, in the early 1800s. **Parentage** is unknown. Chromosomes are 2 x 17 = 34, a **diploid**.

Size is medium large on seedling rootstock. I suggest nothing larger than M.7A rootstock for home use. Soil preference is for heavy soils and does not do well on sandy, leachy soils. If overset, it tends to **biennial bearing** but, with adequate thinning, it is **nearly an annual bearer**. For **winter hardiness** I suggest USDA zones 6 and 7. It usually does not do well farther north in New York, at least upper New York, with fruit of below average size, color and quality.

Quite **disease free** as grown here, but somewhat susceptible to apple scab fungus. **Blooms** in the late-bloom period. **Sources:** most commercial nurseries, especially those in the eastern U.S. Red strains are available.

Fruit:

As grown in zones 6 through 8, fruit **size** is usually medium-large to large if thinned. **Shape** is roundish-oblate, with usually 30 to 50 percent of the fruit with an oblique axis (see drawing.) **Skin color** is red to pinkish-red, often with red stripes. On poorly colored fruit, ground color is greenish to yellow. Some **russet** or scarf-skin may form on either end of the apple. **Stem** length is rather short, but sometimes to the top of the cavity. **Flesh** is yellowish, firm, crisp and slightly coarse, with a tart-sweet flavor.

Storage life is well above average at 30° to 32°F, into March to April. With improper management, it has a **physiological disorder** called cork spot. This is a calcium or boron disorder similar to bitterpit. Too much nitrogen, very low pH, and overpruning contribute to it. Inadequate watering in July, August and September also contributes. **Matures** in late October.

Quality rating: Good to Very good.

❖ ❖ ❖ ❖ ❖ ❖ ❖ ❖ ❖ ❖ ❖ ❖ ❖ ❖ ❖ ❖ ❖ ❖

Some readers may be surprised to find York described and pictured in my current 50 apples, but the reason is simple: we have so few really superior processing apples grown in enough quantity to supply the processing market. York's importance is emphasized by its standing in the top 10 commercial U.S. apples. In the last few years it has been in seventh or eighth place, hovering between six to seven million 40-pound boxes. York has been called the ultimate processing apple for pie and sauce (canned products).

There are, of course, five other processing apples in the top 15, but they also have considerable fresh usage. They are Stayman, Rhode Island Greening, Northern Spy, Gravenstein and Golden Delicious. Gravenstein is processed somewhat in the western states, but, except for Golden Delicious, it is difficult to find trees of the other cultivars out here. Except for the making of apple juice, we have few processing plants in Washington, Oregon and California that can be used for applesauce or apple pie slices. The east coast and Ohio River Valley supply the most apple processing plants.

Today a majority of Yorks are grown in Pennsylvania, Virginia, West Virginia and North Carolina. Beach said in 1903 that the York was not at its best in the state of New York, but was more of a southern apple.

One acquaintance of mine said his family in West Virginia had a special acre of 48 standard (seedling) York trees that averaged just over 50 bushels (40-pound boxes) per tree one year. These were very mature trees with one large old tree giving slightly over 100 bushels. Even allowing for some misfiguring, this would be about 100,000 pounds (2,400 boxes) of apples. They were sold for processing at seven cents a pound. He said they were somewhat biennial, so they only averaged about half that over a 10-year period. He also said that they could live with, on average, a $4,000 gross income per acre. I cannot prove these figures, but York is known as a mortgage lifter in that part of the Virginias.

York's relatively small core gives more apple to process. It is the only commonly grown apple I know of that has some of its fruit, usually about 30 percent on a given tree, that grows with an oblique axis.

When allowed to ripen well and stored a month, it is a firm, fine, fresh eating apple. York is one of the most important commercial apples of the eastern U.S., and it will be well into the 21st century.

Notable New Apples

Goldrush

This new apple is suggested for trial because of its resistance to disease and long keeping qualities. It is field immune to apple scab fungus, very resistant to powdery mildew, somewhat resistant to fireblight, but susceptible to cedar apple rust.

Tested as Co-op #38 at Purdue University in West Lafayette, Indiana. A seedling with parentage of Golden Delicious x Co-op #17. Matures just ahead of Fuji in October, and is reputed to keep four to five months in regular storage of 30° to 32°F. A yellow-color skin that often has a red blush over the yellow. It is slightly tart to taste at picking time. Heavily spurred, it tends to overset and needs thinning to prevent biennial bearing. Blooms in the late season. Fruit is medium in size and shape.

Goldrush has a very good flavor, but I cannot yet rate it. This one may become a real winner in future years. Available from Stark Bros. as of 1995, but I understand other nurseries are being assigned to grow and sell it.

Honeycrisp

Formerly Minnesota #1711 (U.S. plant patent #7197) by the University of Minnesota, a cross of Macoun x Honeygold. It is a large apple with at least 50 percent red color over yellow background. A reliable annual bearer that blooms in early- to mid-season. This is a sleeper and has been acclaimed by testers across the nation, and not just from north of USDA zone 4.

Honeycrisp, like all Minnesota apples, was tested for many years and, as of this writing, may be suggested for USDA zone 3 (northern Minnesota), at least for trial. I suggest testing in zones 4 to 8 and protected zone 3, on cold-hardy root-stock in zones 3 or 4 (not the Malling series). Pick about September 10 to 15 in Minnesota.

Many nurseries have been granted the rights to propagate it; some known to me are Bailey Nurseries, Stark Bros. Nurseries, and Burchell Nursery. Available in 1995-96. In the "Very good" class.

Honeygold

This apple is from the breeding work of the Excelsior Station at the University of Minnesota, a cross of Golden Delicious x Haralson. In certain years, one parent, Golden Delicious, will not ripen in Southern Minnesota and South Dakota. Golden Delicious trees are endangered in a severe test winter or early cold weather. Honeygold usually succeeds in those areas and is a favorite of small direct-market orchardists in southern Minnesota, Wisconsin and South Dakota. This is the yellow apple of those areas, with fine flavor and salability.

The Minnesota apples bred for the north have very catchy names, as does Honeygold. Some are described in this book, such as Sweet Sixteen and Keepsake. If an apple has good qualities, then a catchy name can help make it popular. Available from northern nurseries such as Bailey Nurseries.

Pink Lady™, brand Cripps Pink cultivar

This intriguing name is apparently matched by an unusual apple. It was crossed at the Horticultural Research Center located south of Perth in the very southwestern tip of Australia, surrounded by the Southern Indian Ocean only a few miles away. This area is at 34 degrees south latitude, similar in quantity of sunlight to about Los Angeles in north latitude. The tree first fruited in the late 1970s as a cross of Golden Delicious and Lady Williams, a chance seedling from western Australia. The name does not refer to flesh color (as with Pink Pearl), but refers to its skin color.

Below are most of its favorable traits, with most of the information supplied by E. W. Brandt and Sons of Parker, Washington, who have propagation rights for the U.S.

Responds in low-chill areas, perhaps as low as 400 hours in the mild climate of southwest Australia. There, Pink Lady is a long bloomer (reported as three to four weeks), which probably reflects a climate with not quite enough chill time, in my opinion. If it followed the long blooming pattern here at 45 degrees north latitude, any diploid would cross pollinate it. Because of our long chill time of at least 2,500 hours (Portland, Oregon), it probably won't bloom over such a long period here, and this alone may ripen it a few days earlier than in southwest Australia. A low chill tree, attendant with high quality, means Pink Lady may be a winner in the southern U.S., as well as the coastal U.S. More apple cultivars of high quality coupled with low chill and late ripening are needed. There are very few now. In southwest Australia it ripens at least one week after Granny Smith.

- High tolerance to sunburn.

- A crisp, fine-grained flesh texture that resists browning after being cut.

- High soluble sugars, giving a unique and distinctive taste embracing the good qualities of both parents.

- A long retail shelf life in stores and a long cold storage life.

- Did not develop bitterpit from young, three-year-old trees at the Manjimup Research Center. Needs testing in our climate to verify.

- It has a 50 to 60 percent pink and light red blush on a yellow-green background, unlike most cultivars sold in U.S. retail stores. Very distinctive.

- Size is a uniform medium-large to large.

My thoughts are as follows: When Granny Smith was taken from the south latitudes of 34 degrees to 40 degrees (Australia and New Zealand) to eastern Washington's 45 degrees to 49 degrees north latitudes, the apple received considerably more daily sunlight. To no one's surprise, the time from bloom to ripening was reduced about 15 days. This may also hold true of Pink Lady, which will be tested at our higher latitudes giving more summer sunlight (Washington, Oregon and Southern British Columbia). Testing is the answer, and my first crop will be picked in the late fall of 1995.

Regarding its disease status, it is susceptible to apple scab fungus in my climate.

Rights to propagate and sell trees of Pink Lady in the U.S. belong to Brandt's Fruit Trees, Inc., of Parker, Washington. Finished trees are available in limited quantity for 1995 and 1996. Dave Wilson Nursery of California has also been assigned to sell Pink Lady.

Rubinette

Another new patented variety from Switzerland; this is a gourmet apple with superior flavor and crisp texture in the "Very good to Best" category. It was grown from a seed of Golden Delicious in the Hauenstein Orchard near Rafz. Cox Orange is considered to be the pollen parent.

In shape it resembles a medium-sized Golden Delicious. The stem is longer and thinner than Golden and the stem cavity is shallow. Reddish-orange stripes over a yellow ground color make it resemble the Cox Orange. Some say it is almost a Cox, but more easily grown.

In Washington state and Central Europe, the bloom season is on the early side of mid-season, and normally the fruit picks in the last half of September. The fruit size increases with the tree's age. Average storage time in common cold storage is about 3 months.

Available from Columbia Basin Nursery, Burchell Nursery Inc., Stark Bros.

Sinta

This is a cross of Golden Delicious x Grimes Golden, propagated by the Canadians. When picked ripe, it is an attractive dark yellow, conical-shaped apple. Its quality is "Very good," and it is certainly easier to grow than Grimes. I tested it for a number of years and found it a beautiful and worthwhile apple if properly sprayed.

Its main fault appears to be considerable susceptibility to apple scab fungus and powdery mildew fungus, so it would do best where the springtime is fairly dry and it is sprayed early for control. Available from Rocky Meadow Nursery and Southmeadow Fruit Gardens.

Sundowner™, brand Cripps Red cultivar

This is another apple from the same research station as Pink Lady, south of Perth in western Australia. A cross of Lady Williams x Golden Delicious, with Lady Williams as the seed parent.

It has a number of characteristics similar to its sister Pink Lady, such as low-chill requirements of about 400 hours. This means it probably can be grown successfully in temperate, Mediterranean, and perhaps even some subtropical climates where there is at least 400 hours of chill time. (See "Chill Time.") Also, like Pink Lady, it continues with a progression of flower buds opening over a long period. My personal opinion, not proven, is that a climate with more adequate chill time of at least 600 or 800 hours could shorten bloom periods. Testing in the U.S. is needed, so you will not see this apple in stores very soon, unless they are shipped in from "down under."

Sundowner is smaller than Pink Lady, with most of its apples medium-sized. Flesh is dense, firm and fine-grained. It is somewhat biennial unless chemically thinned. It has a long storage and shelf life and holds flavor in C.A. for about six months. It has great tolerance against sunburn, a compelling sweet-tart flavor, and a streaked red color over a yellow-green background.

Ripening in the southwest Australia area, it is one of the latest, about two weeks later than Granny Smith, and it will no doubt be quite site-selective even in the southern U.S.

A fine new apple waiting for testing with virus-free trees. As with Pink Lady, Brandt's Fruit Trees, Inc. has propagating rights for the U.S., and trees were available for purchase in limited quantities starting in 1995. Also available from Dave Wilson Nursery.

Arkansas Series

Perhaps we apple testers in the West have not paid enough attention to these new, numbered cultivars from the Arkansas series. Some have been named and more are to be named soon, and then perhaps we will aid in testing them in the Pacific Northwest.

New Jersey series (N.J.)

They are all still numbered and being tested except #55 (see Suncrisp below) and 90 have shown promise as they have been taste tested in our area for some time. The question is, are they enough better than current cultivars to warrant competing with orchards of Golden Delicious, Jonagold, Blushing Golden™, etc.? The apple tree and its ease of growing in various climates is almost as important as good flavor.

In other words, the New Jersey series has not yet been tested enough in the west. Maybe 10 years from now, some will emerge as really worthwhile.

Suncrisp

This is the selection formerly called N.J. #55, a selection of the New Jersey breeding program. Its two-generation genealogy is as follows:

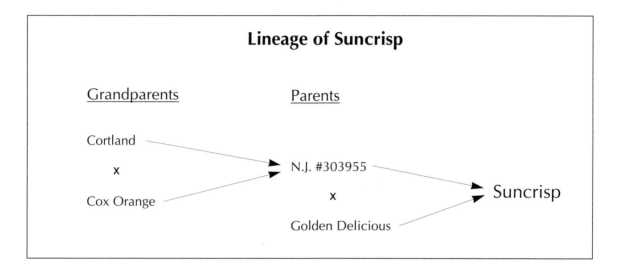

Lineage of Suncrisp

Grandparents   Parents

Cortland
x → N.J. #303955
Cox Orange   x → Suncrisp
Golden Delicious

Its skin color is greenish to yellow, frequently with an orange-red blush. It matures about one week after Golden Delicious and stores about three to five months in common cold storage (30° to 32°F.)

This apple is available as of this writing with a testing agreement from Rutgers Fruit Research and Development Center.

This is for the curious tester who can get a testing agreement. Also, a plant patent has been applied for and trees will be available from Stark Bros. Nursery by 1995. Other nurseries will also get growing and selling permission.

After tasting this fruit over a three-year period, I would rate it in the "Very good" class.

Series from Summerland Research Station in British Columbia

Dr. David Lane and his colleagues have made rather large quantities of apple crosses in the last 10 to 20 years; some are quite worthwhile, and one is named.

It was my privilege to taste many of these at the Pacific Northwest Testers Association meeting in December 1990 at Yakima, Washington. Three or so were "Very good" in flavor, but the other characteristics of the trees and apples are unknown to me.

Testing is just starting outside of the Summerland, B.C., Station, to determine which might have commercial value. Many should have backyard value. One was #8C27-96, soon to be called Sunrise, which has "Very good" flavor for a mid-August apple.

This numbered series has some above-average specimens that may stand the test of time in different climates, especially the Gala x Splendor crosses. Here again, as yet, I cannot write much about them in a definitive manner.

So, all over the world we are beginning to test at least 150 above-average new apple cultivars, but it is too soon to determine which will emerge as acceptable out of these dozens. Most will not!

I wonder how many new apple cultivars will be accepted by home orchardists, let alone commercial growers? Commercial apple commissions of certain states have been quoted as saying that in the last 10 years we have had about eight fine new apple cultivars being planted from coast to coast; we don't want anymore, at least not now. I have noticed that many commercial growers act threatened by this constant parade. We all resist change, especially if it costs quite a bit of money to make these changes.

Yet the rush to find a number of above-average apples goes on, and it does produce scientific background information as reported by tester associations, especially finding which cultivars grow best in our various climates.

Of course constant testing must continue, as it has for over a century, or we might still be eating tasteless Ben Davis and struggling with seedling rootstock and huge, 30-foot trees. It is all a journey, not a destination.

Part III

Bloom Periods

Earliest Bloom

Erwin Baur
Gravenstein
Idared
Grimes Golden
Dolgo Crab
Whitney Crab
Braeburn, in New
    Zealand and much of
    lower midwestern U.S.

Early to Mid Bloom

Liberty
Empire
Mutsu
Young America Crab
Rhode Island Greening
Ashmead
Arlet
Cortland
Haralred
Spitzenberg
N.Y. #429
Wickson Crab
Hudson Golden Gem

Mid to Late Bloom

Jonathan
Winesap
Stayman
Keepsake
Sweet Sixteen
Blushing Golden™
Bramley
Braeburn, in much of
    Washington and Oregon
Gala
Cox Orange
Karmijn de Sonnaville™
McIntosh
Brock
Jonagold
Freyberg
Kidd's Orange Red
Ginger Gold™
Sturmer
Roxbury Russet
Golden Delicious
Red Delicious
Newtown
Melrose
Fuji
Elstar
Tydeman's Late Orange
Coromandel Red

Late Bloom

Granny Smith
Calville Blanc d'Hiver
York
Spigold
Northern Spy

Very Late Bloom

Orleans

My Test Orchard

The orchard dates on which flowers were 80 percent to 90 percent open in two different years, on the same trees, were:

	Very Early to Mid Bloom	Mid to Late Bloom	Late Bloom
1986	3/28 to 4/6	4/9 to 4/16	4/15 to 4/27
1989	4/12 to 4/18	4/16 to 4/23	4/25 to 5/6

Flowering was about 10 days later in 1989 than in 1986 for all the bloom periods. This is one reason I have not used dates as a guide.

Bloom periods are provided because they more accurately reflect compatible pollenizing cultivars. The bloom dates will vary by up to two months between USDA zone 3 and zone 9, so dates sometimes seem of small value even for a given area. Higher altitude will also delay bloom.

Keep in mind that in northern areas all bloom periods will have considerable overlap when spring comes suddenly and stays. As long as chill time of 32° to 45°F have been met for each cultivar, the cool or cold early spring will hold them all from blooming until the real spring comes suddenly and warms quickly.

Please note the great variation of Braeburn blooming periods, as stated below:

In New Zealand, southern Illinois and southern Indiana, Braeburn is reported to bloom with, or before, the early blooming Gravenstein. Here in the Willamette Valley of Oregon and much of southeastern Washington, it has proven to be a mid- to late-bloomer.

I cannot account for this except that our usual weather from about March 15 to the end of April is cool, cloudy and drizzly and holds back bloom. Our chill time (32° to 45°F) has been met. Its bloom periods have Jekyll-and-Hyde overtones; the apples look and taste alike, so we appear to have the same apple. Later blooming means later maturity (ripening.)

This is just one reason why a bloom chart such as this should be made for your own area, although most cultivars do follow a trend. Except for the Rosebrook strain of Gravenstein, Gravenstein appears to be an early bloomer anywhere, as it should be.

Apple Shapes

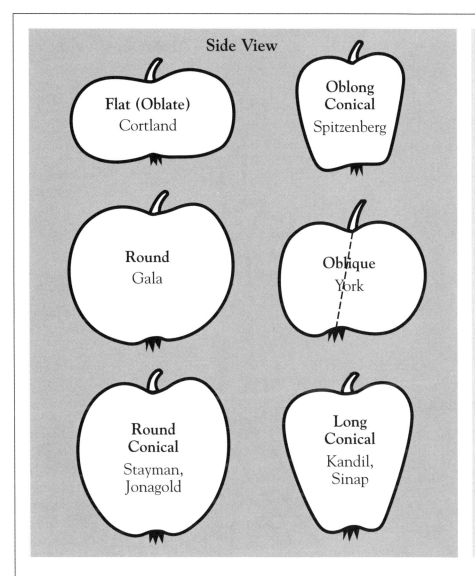

Side View

Flat (Oblate)
Cortland

Oblong Conical
Spitzenberg

Round
Gala

Oblique
York

Round Conical
Stayman, Jonagold

Long Conical
Kandil, Sinap

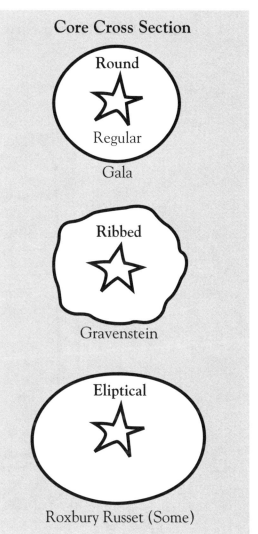

Core Cross Section

Round
Regular
Gala

Ribbed
Gravenstein

Eliptical
Roxbury Russet (Some)

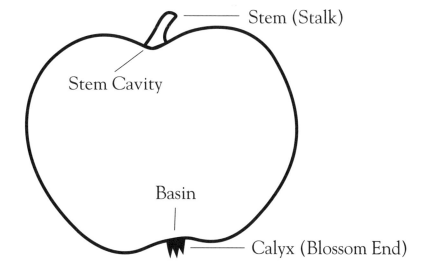

Stem (Stalk)

Stem Cavity

Basin

Calyx (Blossom End)

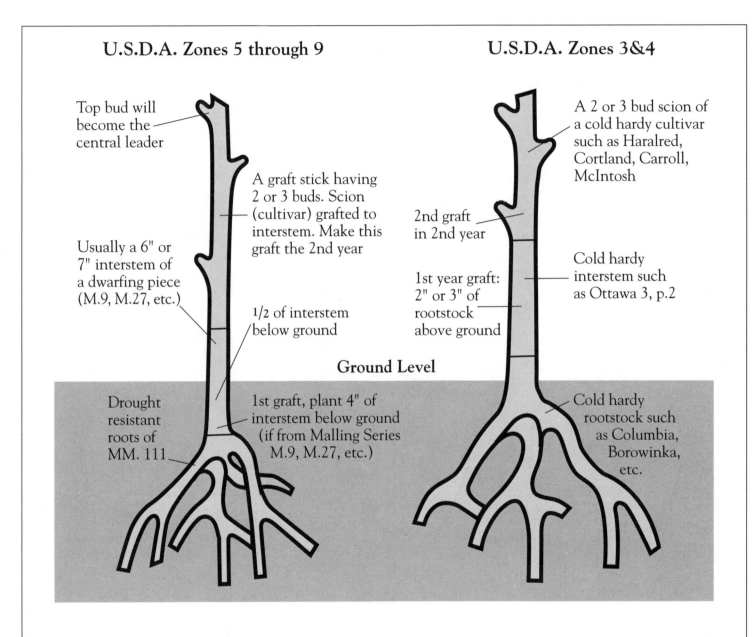

U.S.D.A. Zones 5 through 9

Top bud will become the central leader

A graft stick having 2 or 3 buds. Scion (cultivar) grafted to interstem. Make this graft the 2nd year

Usually a 6" or 7" interstem of a dwarfing piece (M.9, M.27, etc.)

1/2 of interstem below ground

Drought resistant roots of MM. 111

1st graft, plant 4" of interstem below ground (if from Malling Series M.9, M.27, etc.)

U.S.D.A. Zones 3&4

A 2 or 3 bud scion of a cold hardy cultivar such as Haralred, Cortland, Carroll, McIntosh

2nd graft in 2nd year

1st year graft: 2" or 3" of rootstock above ground

Cold hardy interstem such as Ottawa 3, p.2

Cold hardy rootstock such as Columbia, Borowinka, etc.

Ground Level

The scion grafting stick should have at least 2 buds. Some like 3 or 4. The top bud should should be about 1/8" to 1/4" from the cut-off point, and it will become the central leader. It should be kept growing straight up for proper hormone flow– so the young tree will grow up to 5-7' the first 2 or 3 years. In the spring (April as a rule), the interstem may be 2' to 5' tall, and must be cut off at the proper length– usually 6" or 7" long– before the 2nd graft of the cultivar is attached.

Apple Crops by States and Regions
USDA statistics (42-pound boxes)

State/region	5-year average	1987	1988	1989
New York	23,762,000	20,952,00	21,190,000	22,619,000
Pennsylvania	13,048,000	10,950,000	11,905,000	9,524,000
Virginia	10,743,000	11,452,000	11,429,000	9,286,000
North Carolina	7,429,000	9,286,000	8,929,000	4,762,000
West Virginia	5,167,000	4,286,000	5,119,000	3,810,000
Massachusetts	2,257,000	2,286,000	2,357,000	2,190,000
New Jersey	2,357,000	1,905,000	1,667,000	1,071,000
Maine	1,919,000	1,786,000	2,238,000	1,667,000
Maryland	1,700,000	952,000	1,048,000	1,357,000
New Hampshire	1,243,000	1,190,000	1,357,000	1,333,000
Vermont	1,100,000	1,048,000	1,024,000	1,070,000
Connecticut	1,052,000	1,072,000	1,167,000	1,071,000
Georgia	809,000	1,190,000	833,000	714,000
South Carolina	733,000	1,072,000	952,000	857,000
Delaware	547,000	619,000	595,000	357,000
Rhode Island	117,000	119,000	131,000	120,000
Eastern U.S. totals	73,983,000	70,165,000	71,941,000	61,808,000
Michigan	20,810,000	25,000,000	19,048,000	23,810,000
Ohio	2,952,000	3,571,000	2,262,000	2,976,000
Illinois	2,281,000	2,452,000	2,024,000	2,214,000
Indiana	1,448,000	1,714,000	1,333,000	1,524,000
Wisconsin	1,410,000	1,548,000	1,071,000	1,405,000
Missouri	1,128,000	1,262,000	1,333,000	1,262,000
Minnesota	500,000	619,000	332,000	524,000
Kentucky	352,000	500,000	262,000	405,000
Tennessee	269,000	357,000	298,000	274,000
Kansas	231,000	286,000	285,000	214,000
Iowa	221,000	238,000	202,000	286,000
Arkansas	252,000	95,000	238,000	214,000
Central U.S. totals	31,854,000	37,642,000	28,688,000	35,108,000
Washington	76,262,000	119,048,000	88,095,000	107,143,000
California	13,167,000	15,476,000	13,095,000	15,476,000
Oregon	3,619,000	5,000,000	3,929,000	4,286,000
Idaho	3,033,000	3,690,000	3,214,000	3,452,000
Colorado	1,919,000	2,976,000	1,667,000	1,786,000
Utah	1,248,000	1,619,000	952,000	1,548,000
New Mexico	203,000	300,000	262,000	214,000
Western U.S. totals	99,451,000	148,109,000	111,214,000	133,905,000
U.S. totals	205,288,000	255,916,000	211,843,000	230,821,000

Commercial U.S. Production by Cultivars
USDA statistics (42-pound boxes)

Cultivar	1987	1988	1989
Red Delicious	114,940,000	88,570,000	97,180,000
Golden Delicious	41,370,000	36,560,000	36,370,000
McIntosh	16,500,000	15,380,000	15,980,000
Granny Smith	10,550,000	12,010,000	14,250,000
Rome	15,140,000	13,900,000	13,510,000
Jonathan	9,570,000	8,320,000	8,790,000
York	6,800,000	7,000,000	5,750,000
Stayman	5,220,000	4,710,000	4,410,000
Newtown	4,250,000	3,930,000	4,150,000
Cortland	3,120,000	2,550,000	2,790,000
R.I. Greening	2,730,000	2,250,000	3,420,000
Winesap	4,070,000	3,520,000	3,630,000
Idared	3,440,000	3,350,000	4,160,000
Northern Spy	3,080,000	2,400,000	2,730,000
Gravenstein	2,550,000	1,850,000	2,140,000
All others	12,588,000	11,736,000	12,442,000
Grand total	255,918,000	218,036,000	231,702,000

Note: Sum of varieties may not add to grand total due to rounding of individual varieties.

Since found, the average age of our top 15 commercial apples (listed above) is about 170 years. So many people say "Why don't we grow some of the old apples?" Perhaps they mean "why don't we grow more of the old apples?" The answer is that commercial growers will only grow a few. Beyond that, you, or a local direct marketer, will have to grow others.

Production Figures for 1992-1993
from USDA figures and estimates
(in millions of 42-pound boxes)

Cultivar	1992	1993
Red Delicious	108,690,000	108,070,000
Golden Delicious	39,060,000	37,450,000
Granny Smith	16,830,000	16,370,000
Rome	15,230,000	16,230,000
McIntosh	16,810,000	14,730,000
Jonathan	9,160,000	8,120,000
York	6,720,000	6,720,000
Idared	5,060,000	5,060,000
Fuji (estimated)	N/A	4,580,000
Gala (estimated)	N/A	4,170,000

Top 10 States by Production
from USDA records
(estimated for 1993)

State	1992	1993
Washington	114,286,000	114,286,000
Michigan	25,714,000	26,190,000
New York	27,857,000	24,286,000
California	20,000,000	20,238,000
Pennsylvania	11,905,000	13,095,000
Virginia	8,810,000	9,048,000
North Carolina	5,714,000	7,619,000
West Virginia	5,357,000	5,119,000
Oregon	4,167,000	3,690,000
Idaho	1,786,000	3,571,000

30 Important Commercial Apple Cultivars and Their Origins

(also suitable for the connoisseur or domestic orchard)

Arlet, Switzerland
* Blushing Golden™, Illinois
* Braeburn, New Zealand
Brock, University of Maine
Cortland, New York
Cox's Orange, England
Delicious (now Red Delicious,) Iowa
* Elstar (Lustre Elstar), The Netherlands
* Empire, New York
* Fuji, Japan
* Gala, New Zealand
* Ginger Gold™, Virginia
Golden Delicious, West Virginia
Granny Smith, Australia
Gravenstein, Italy or Denmark
Haralred, University of Minnesota
Idared, Idaho Experiment Station
* Jonagold, New York
Jonathan, New York
* Liberty, New York
McIntosh, Ontario, Canada
Melrose, Ohio State University
Mutsu (Crispin), Japan
Newtown, Long Island, New York
Northern Spy, New York
Rhode Island Greening, Rhode Island
Stayman, Kansas
* Sweet Sixteen, University of Minnesota
Winesap, New Jersey
York, Pennsylvania

* New apples with much promise for the next 20 or more years.

Most of these cultivars tend to biennial bearing so should be thinned at blossom time before hormone flow starts. Northern Spy, York, Rhode Island Greening, Gravenstein, Golden Delicious, Sweet Sixteen, Braeburn and Fuji should be thinned as early as possible.

Virtually all cultivars need some thinning, but some require hormone or chemical thinning at blossom time or they will produce little or no crop the following year.

Apples Primarily for Connoisseur and Domestic Orchards

Most of the cultivars listed below should be climate tested before being planted in quantity for commercial growing. Kidd's Orange Red, Bramley, Sturmer, Coromandel Red, New York #429, Spigold, Tydeman's Late Orange, and Keepsake are being grown commercially in a limited way and all are of outstanding flavor in the correct environment. Except for direct marketers, these cultivars receive scant attention in the U.S. from larger commercial growers using packer-sales outlets.

Ashmead, England
Bramley, England
Calville Blanc D'Hiver, France
Coromandel Red, New Zealand
Erwin Baur, Germany
Freyberg, New Zealand
Grimes Golden, West Virginia
Hudson Golden Gem, Oregon
Kandil Sinap, Turkey
Karmijn de Sonnaville™ (Karmine), The Netherlands
Keepsake, University of Minnesota
Kidd's Orange Red, New Zealand
New York #429, Geneva, New York
Orleans, probably France
Roxbury Russet, Massachusetts
Spigold, Geneva, New York
Spitzenberg, New York
Sturmer, England
Tydeman's Late Orange, England[*]
Wickson Crab, Ettersberg, California

[*]Tydeman's Late Orange, a very superior apple, is three-quarters Cox Orange without the growing difficulties associated with Cox. It must be thinned early and heavily or it bears biennially. Do not confuse this with Tydeman's Early (Tydeman's Red), a McIntosh cross of rather average quality. Others need thinning also, as most apples are not annual bearers without some help.

Part IV

Apple and Apple Rootstock History

When humans first paid attention to apples as food is lost in the antiquity of pre-history; it was before the Romans, the Greeks and their ancestors. Any written records we have are mostly from the Greeks and Romans.

Theophrastus was a Greek botanist and philosopher, and in 300 B.C. he indicated that Alexander the Great had brought dwarfing apple trees (or seeds) from Asia Minor (Turkey, Russian Caucasus Mountains, Iran). His death in 287 B.C. kept us from a better historical record of his time.

History shows the Romans conquered Greece and brought their rootstocks and perhaps cultivars to be grafted back to Rome. Both these civilizations probably learned grafting from settlements farther east, so this art is also very old. Then Rome moved north for further conquests, and they brought apple seeds as well as grafting techniques into France. Some of the very old apples such as Lady (Api) and Court Pendu Plat may have come from that period.

Some old books have said that by 100 A.D. the Romans had named many apple cultivars; but, of course, we cannot prove that and refer only to old reports such as from Pliny the Elder (23-79 A.D.) who perished in the eruption of Mt. Vesuvius that buried the City of Pompeii (he was reputed to be aboard a ship that capsized).

By the 1500s to 1800s, apples were grown on two general types of dwarfing rootstocks in France. The largest was called Doucin and probably gave a number of like-size trees, but were not clonal. The other was Paradise from Asia Minor, which gave smaller trees. Probably the Yellow Metz (M.9) of France came from one of the Paradise trees. The point is, these dwarfing rootstocks giving various-size trees are not new. The mixing of these size-controlling and/or early fruiting rootstocks became so confusing that standardization was started in Germany, Belgium, Netherlands and France in the late 1800s.

In 1912, R. Wellington began a study of the Paradise forms at the Wye College Fruit Experiment Station in Kent, England. This station was later named the East Malling Research Station. When Wellington joined the armed forces, his work was assigned to R. G. Hatton, just before the fateful days of August 1914 when Britain entered World War I. (Compact: International Dwarf Fruit Tree Association.)

By 1917 Lord Hatton had numbered nine types of rootstock. By 1929 to 1935 he or Dr. Tydeman, who continued Hatton's research, had numbered 26 types, and they used Roman numerals; i.e., EM.I to EM.XXVI. Later, Arabic numbers were used instead of Roman numerals, i.e., EM.1 to EM.26. The initials "EM" refer to East Malling (East Malling Research Station). This has been further simplified to M.1 through M.26.

A few of the original 26 rootstock variety selections were already several centuries old when they were numbered (such as M.7). Some Malling rootstocks have not yet been widely tested in the U.S. because many of them are not size-controlling at all; rather, they have early fruiting habits, cold hardiness, etc. The rootstocks that limit tree size are the ones in most demand.

All the rootstocks on the listed trellis chart require some type of support, such as a three- to five-wire trellis, with any cultivar grafted to them. This is a modified "French Axe" system. A heavy two- to three-inch, rot-resistant wooden stake can also be used, but it is much more difficult to use when tying branches down on young trees and, after fruiting, to tie heavily laden branches from hanging straight down. I prefer the trellis for home or small commercial orchard use. M.7A rooted spur strains do not require support except in sandy, windy sites, but I like doing it to keep a tree straight. Such support for all these on the trellis chart is for 15 years at least, and perhaps for the life of the orchard.

Rootstock in America

To beginners in fruit tree growing, specifically apples, it is often confusing that we do not plant a seed of a Jonagold to get a Jonagold, or a seed of any cultivar we want to reproduce. They do not breed true and, as with people, offspring may resemble parents but they are not the same. An apple's seeds often produce inedible apples, all different. On the other hand, we can only get a new variety by planting seeds–thousands of them–and, occasionally, an above average new apple comes from one of them. It is a chancy long-term program to get new varieties.

To get another Jonagold, one needs to graft a bud stick (2 to 4 buds) from a Jonagold tree onto a rootstock, or insert one bud on a rootstock. This is the way to get the same cultivar reproduced, and the only way the same cultivar can be reproduced genetically for hundreds and even thousands of years.

Rootstock has become increasingly important because there are so many different ones for different climates and soils. Some are cold hardy, some need support because of brittle roots, some give small trees, some give large trees, etc. We need to know the differences, and testing them takes many years.

When I was a boy, all of our rootstocks were seedling, many from cold-hardy crabs. Since each seed had its own genetics, they did not all give like trees with the same cultivar grafted to it; however, these seedling trees (standard) were often long-lived and usually hardy into USDA zone 4. Those that were not hardy in that zone died. In the 1930s, the main tree size difference came from the cultivar. A Northern Spy gave a large tree on seedling rootstock. Haralson or Wealthy are genetically dwarfing and gave much smaller trees on seedling rootstock.

The first settlements in North and South America were in the tropics or subtropics, primarily by the Spanish and Portuguese, and they brought a citrus culture, not apples.

Starting in the early 1600s, the northeastern United States and Canada were settled primarily by northern Europeans, and they brought large quantities of apple seeds, as well as grafting wood of their favorite cultivars. For about 300 years in America, millions of apple seeds were planted as frontiers moved west. Seeds were used because of lack of rootstock. There was very little grafting wood of known cultivars available, and many pioneers did not know how to graft anyhow.

John Chapman (1775-1847), better known as Johnny Appleseed, was an example of one who planted great numbers of apple seeds. There is evidence that many of these orchards were later grafted to known cultivars, instead of putting up with worthless seedlings; however, he helped to get many orchards started in the early 1800s. Whether any of these seedlings were superior and survived as known cultivars is unknown.

Probably never before were conditions so perfect for the planting of enormous numbers of apple seeds as in pioneer America, and about 5,000 to 6,000 cultivars were named, with perhaps 200 or so still being reproduced. The less edible seedling apples were used for animal food, and supplied the farms with the main drink of pioneers, hard apple cider of about five to seven percent alcohol, and apple juice. A few seedlings, such as Newtown and Spitzenberg, were superior and survive today.

Today, the amateur planting seeds in his garden or orchard to find a new variety is virtually non-existent. Instead, we can and do graft cultivars we like, or buy such a tree already grafted. Research stations do some breeding work by planting seeds of known parentage, so new cultivars are still arising. Fortunately, there is an international exchange of results. New Zealand and Australia have supplied us with Braeburn, Gala, Freyberg, Granny Smith, and now Pink Lady and Sundowner. Europe has given us Cox Orange, Elstar, Arlet, and others. We, in turn, have supplied them with Golden Delicious, Jonagold, Jonathan, Idared, Red Delicious and others.

The Malling rootstocks have served us well, but the newer Cornell-Geneva rootstocks designed for North American diseases, soils and climates will soon be available for test orchards. These newer rootstocks surely will help us conquer or control such problems as fireblight, collar rot, voles (large, short-tailed rodents), and cold climates as much of the Malling series has not done. Even the Polish (P) and Budagoski (B) rootstocks from Poland and Russia, which were developed for cold hardiness, were not developed to resist fireblight because it was not a problem in northern Europe then (40 to 50 years ago). In the last 15 years, fireblight has swept across much of Europe. Only southern France and Italy are free of it still, and probably, they will not be much longer. Newer rootstocks then should have resistance to this serious bacterial disease.

This book cannot cover all of apple and rootstock culture, as it would require three to four volumes and another lifetime of experience; however, I wanted to give home orchardists a start, so that the grower might only make the minor mistakes which are all part of the learning process. If one does not have some basic knowledge, then growing superior quality apples can be difficult, especially in moist, warm day and night climates such as the lower midwest and below the Mason-Dixon line. Eastern Washington, with its dry climate, hot days and cool nights is ideal for the Red Delicious strains, so 55 percent of the nation's apples come from that near mono-culture. Now this area is slowly adding other cultivars suited to its climate, such as Fuji, Gala and some Braeburn. Jonagold and Elstar are not suited for this hot area.

Rootstocks are the most important factors we need to learn about. If they are not correct for one's climate and soils, a futile battle ensues that often means the removal of trees. The cultivar can be changed by top-work grafting if it is wrong and the rootstock is right.

I am most anxious to test the Cornell-Geneva series which was developed for North America and am delighted that one local nursery, Meadow Lake Nursery of McMinnville, Oregon, will start sales of these rootstocks in 1995-1996. Some grafted trees should be available by 1997.

Dr. Jim Cummins and his associates used various disease resistant and cold-hardy crabs in their crosses for better rootstock. Albert Etter made such wide crosses for apples with crabs in the early 1900s but, as a modern-day Luther Burbank, he was not part of the "establishment" and the establishment was not listening. Fortunately, Dr. Cummins and his associates are some of our best researchers. I have hopes his 30 years of work with new rootstock at Cornell-Geneva will be of great aid to North American growers.

The Malling series of rootstock is fine for my climate in the Willamette Valley, which is similar to that of southern England, especially since we do not have fireblight to contend with. Few of these Malling rootstocks that give smaller trees are cold hardy in USDA zones 3 or 4, but that also is not a problem here. Malling rootstocks simply fail in USDA zones 2, 3 and 4, and many are susceptible to fireblight and collar rot.

Thomas Jefferson has always been my ideal as a fruit grower and as a president. Even though he had superior knowledge crossing a number of disciplines, he often stated that growing food was a main interest, and experimenting with cultivars from everywhere was important. (His attempt to introduce olive trees in South Carolina is one example.) Jefferson, Madison, and Washington were good agriculturists, and men of culture and wealth emulated them. They, especially Jefferson, believed practice and practical study were paramount to success. It is unfortunate that Jefferson had little opportunity to use dwarfing rootstocks, or the use of them in America would probably not have been delayed so long.

Every 25 to 50 years a better rootstock will probably be developed for an area, but I cannot guarantee that the grower will not have local problems because of the many variables involved. Growing superior fruit requires study and close observation by the grower, but this does not make it a less fascinating pastime.

Trellis Chart Rootstock

The purpose of this section is to acquaint you with the usual characteristics of these rootstocks and the tree size you can expect from them. The cultivar also has a strong effect on tree size, so there are four different size groups for each rootstock with a particular cultivar.

There are many rootstock nurseries, especially in the eastern U.S., that are unknown to me. It is not one of the purposes of this book to research dozens of nurseries and what they do, because so many changes and additions in the apple rootstock field occur frequently. Those mentioned will give the reader and fruit societies a starting point for locating apple rootstock.

Most dwarfing

P.22: USDA zones 4 through 8. Possible in zone 3.

This is a new rootstock developed by the Research Institute of Pomology, Skierniewice, Poland. It is a cross of M.9 and an old cold-hardy Russian apple, Antonovka.

P.22 is quite resistant to most disease, but not to fireblight or the insect woolly aphid. It is a fine rootstock in my climate, but produces a miniature tree.

P.22 is as small as EMLA.27 but is much more cold-hardy and can be used into USDA Zone 4, where the Malling series usually fails in test winters.

Some believe P.22 is too small for commercial production unless one is willing to buy at least 800 to 900 trees per acre and plant every 4 feet to 5 feet within rows and 8 feet to 10 feet between rows.

P.22 is also being used as an interstem (three-piece tree) on cold-hardy rootstock. It will require support in its early years.

<u>M.27 (originally EM3431): I suggest USDA zones 6 through 8 and protected zone 5.</u>

Moderate tree size cultivars grafted to M.27 rootstock are quite small and can be held at 6 feet to 8 feet or less. It must be staked or trellised because of brittle roots.

This rootstock is presently only accepted in North America by home orchardists who like its good production and very manageable size. It grows well here in USDA zones 7 and 8. Reports from upper midwest universities in USDA zones 3 and 4 have shown a lack of cold hardiness in test winters. For very small trees in colder zones, P.22 should probably be used.

My experience with M.27 suggests one should push it with some fertilizer the first two or three years to get it quickly to 5 to 7 feet. Once it starts fruiting heavily it tends to runt out more than any rootstock I have tested. It grows very slowly when heavily fruiting during the third or fourth leaf and requires little or no pruning.

It is used in the lower midwest (zones 6 or 7) as an interstem on MM.111 rootstock. This produces a free-standing tree about the size of trees rooted on M.26, with good drought resistance from the MM.111 rootstock.

Dwarf group

<u>M.9: USDA zones 6 through 8 (and protected zone 5).</u>

This is an old rootstock, part of the original Malling series. It was first known in France shortly after 1870 as a chance seedling, probably of the very ancient Paradise. In the last 70 years, many strains have appeared in Europe and at least three in the U.S. I do not know their differences.

The chaotic situation regarding M.9 strains hopefully will be sorted out, and the M.9 designation will include a strain number or name, such as M9 (#37). We will then know better what we have planted.

Because of the uncertain characteristics of the many new M.9 clones, I have planted trees almost exclusively with the old original M.9. Despite containing latent virus it has served well for over 100 years. Indeed, the two or three latent viruses in this old rootstock are one reason why it is so small and slow growing. Such rooted trees can be held to 8 or 10 feet without rampant regrowth. I like it better than any rootstock tested in my orchard, for my climate and soil. Other growers prefer virus-tested stock.

217

Because unsupported M.9 tends to fall over with its first heavy fruit load, many have assumed it has a shallow root structure. I dug some up about 20 years ago and found root tendrils four to five feet down in deep clay-loam soil. The problem is not shallow roots but extreme brittleness in the roots. As a result, these trees need a three to four wire trellis for the life of the orchard. I prefer the trellis so that heavily laden limbs can be supported.

M.9 strains are the predominant rootstock today in southern England, northern France, Belgium and The Netherlands. These are all areas near large bodies of water and have maritime climates similar to the Pacific Northwest, especially here west of the Cascade Mountains. Unless heavily mulched and with snow cover, it has not survived in zone 4 at the University of Minnesota. In severe test winters in USDA zone 5 in the midwest it can also fail.

M.9 is quite resistant to collar rot on the roots, which occurs at and just below ground level. It is susceptible, however, to a bacterial root gall called crown gall (*Agrobacterium tumifaciens*) when planted in infected soils. It is also susceptible to fireblight, and susceptible cultivars such as Jonathan and Idared probably should not be grafted to this rootstock in bad fireblight areas; better yet, use a more resistant rootstock.

M.9 is somewhat intolerant of soil temperatures over 70°F, and I suggest mulching in hot climates, or anywhere. Suckering is sometimes a problem. M.9 rooted trees bloom profusely and crop heavily, which contributes to their slow growth.

P.2: USDA zones 3 through 9.

This is another cold-hardy rootstock from the series propagated at Skierniewice, Poland, and it is also a cross of M.9 and Antonovka. Its size is about the same as cultivars on M.9 rootstock. In USDA zones 3 or 4 it can be used instead of the Malling series for trees of this size.

P.2 is quite resistant to crown gall, various cankers, and collar rot. It is as susceptible to fireblight as M.9 but does not sucker. It is reputed to be free-standing, but I would still want a trellis.

P.2 induces early defoliation of the cultivar and late bud break in the spring. Both of these characteristics contribute to cold hardiness. It also is being tested as a promising interstem on cold-hardy rootstocks.

<u>B.9: I suggest USDA zones 4 through 8.</u>

B.9 comes from the research of Dr. Budagovski, who was probably the Soviet Union's most successful apple and apple rootstock breeder. He died in 1975 after 40 years in the field of horticulture, and worked at the Michurinsk College of Agriculture where temperatures often reached -25°F.

B.9 produces trees marginally larger than M.9. It is from a cross of M.8 (Malling series) and Red Standard, a hardy red-leafed Russian apple clone. It is very resistant to collar rot and moderately resistant to apple scab fungus and powdery mildew fungus. Like M.9, it is susceptible to fireblight.

B.9 is a replacement for M.9 in the USDA zones where M.9 fails. It requires support as M.9 does. I think I can sum up B.9 by saying that it is almost a duplicate of M.9 except that it is cold-hardy one to two USDA zones farther north.

Semi-dwarf group

<u>EMLA.9: USDA zones 6 through 9 (and protected zone 5).</u>

The differences between this strain and its M.9 predecessor are sufficiently important to me to warrant some explanation. EMLA (East Malling-Long Ashton) means that this clone went through heat treatment for virus removal. Even before heat treatment EMLA9 was a larger clone than the original M.9, and after virus removal it became larger still.

There is some evidence that when Hatton and his colleagues collected rootstock from France, he called all the M.9 physiological look-alikes (in leaf structure, etc.) M.9, even though there was probably more than one clone of Yellow Metz already in existence. (Compact: International Fruit Tree.)

In any event, EMLA.9 rootstock produces a tree larger than my old latent virus infected M.9. In our fertile soils, warm summer climate and with irrigation, it seems to give trees nearer in size to M.26 than to the old M.9. This may not be the finding of others who have tested both in more stressful climates and weaker soils than mine.

The current EMLA.9 often does not root well on one side but, since it needs staking or trellis support (as do all M.9 clones), this won't make a great deal of difference. Even though nearer MARK (see below) or M.26 (see below) in size, it is not susceptible to collar rot as is M.26. EMLA.9 has high productivity.

<u>MARK, formerly called MAC.9 (Michigan apple clone), Michigan State University plant patent no. 091079: USDA zones 5 through 8 (and protected zone 4.)</u>

I did not find that my trees on this root would stand without support. They tended to grow trees that were crooked and leaned. I put in a two-wire trellis and will leave it in until my six-year-old trees are at least another two years old. The roots may produce large etiolation swelling which makes the passage of liquids difficult and slow, especially in light or sandy soils. This swelling is causing failure in many areas due to trees drying up or dying. Some northern areas with irrigation are finding no problem with this.

Its main advantages over M.9 are no suckering, stronger anchorage, and better cold hardiness. In my estimation, this rootstock will probably no longer be used in a few years, except in heavier soils and with irrigation. It is definitely not for areas with sandy soils and spasmodic rainfall.

<u>M.26 (Synonym, EM3436): I suggest USDA zones 4 through 8.</u>

This rootstock was derived from a cross of M.9 with M.16. M.16 had been named the Ketziner Ideal in Germany and was known to be very cold-hardy. This cold-hardy trait persisted in M.26. M.26 is the only commonly grown Malling rootstock to survive the frequently severe test winters at the University of Minnesota, such as the one in 1983-84, but it does have its limits.

M.26 produces a tree that is 15 to 20 percent larger than those on M.9 in the same soils and climates. One must be careful with certain vigorous triploid cultivars on this root as they become shockingly large, especially Gravenstein, Mutsu, Baldwin and Holstein. On the other hand, when the heavily fruiting triploid Jonagold is grafted to M.26, the tree size is medium, but I use minimum nitrogen and bend the leader over to restrict size.

M.26 rootstock should not be free-standing the first five to seven years, or too many trees fall over or lean badly, *unless* one is willing to defruit them for some years and keep them growing very straight. Because M.26 roots tend to make certain cultivars grow in a crooked configuration, it is difficult to keep them growing straight during any time frame. Most of the small commercial orchardists I work with prefer to use the trellis.

Disease resistance varies. In eastern North America M.26 is sensitive to tomato ring-spot virus, which is transmitted in soils infested with the dagger nematode. The virus affects the graft union, usually killing the tree, and is called brown-line decay or apple union necrosis. In poorly drained soils that occasionally puddle, it is susceptible to collar rot (*Phytopthera cactorum*) and other *Phytopthera* species. Although it does not sucker, the root is susceptible to fireblight. In those areas with this bacterial disease, M.26 rootstock should not be used, and M.7A is a better choice.

Cultivars on M.26 rootstock are quite susceptible to drought and I would not use it in *any* areas depending on variable summer rainfall, unless supplementary irrigation has been supplied.

Here in the Willamette Valley we summer irrigate almost all apple rootstocks so M.26 has proven to be an outstanding performer, giving very high production. It seems easy to care for on a trellis as we do not have fireblight to worry about, and apple union necrosis has not been detected in this area. EMLA.26 is a virus-free clone.

Ottowa.3: USDA zones 3 through 8.

This is one result of clonal rootstock work done in Ottowa, Canada, and resulted from a cross of Robin Crab and M.9. Robin Crab is very cold-hardy and imparted this characteristic to Ottowa.3. It has some susceptibility to fireblight and woolly aphids, a slight tendency to sucker, some susceptibility to brown line decay and is difficult to propagate.

In spite of these faults, Canadians feel it is a promising stock for the north. Quebec researchers have found it the best cold-hardy *interstem* to dwarf the rootstock of the very old cold-hardy Russian apple Beautiful Arcade. This combination may become an important north country plant to graft, or bud, cold-hardy apple cultivars to, where the Malling series has failed. Most nurseries resist doing interstems because of the additional time and cost.

Interstem

M.9/MM.111: USDA zones 5 through 8, suggested.

Under certain growing circumstances the three-piece trees are a marvelous invention. Are they new? Hardly! They were written about hundreds of years ago.

The three pieces are the vigorous drought-resistant MM.111 rootstock with a seven or eight inch interstem of M.9 grafted to it, which dwarfs the tree, and the top piece is the scion of a specific apple cultivar. It is best to do this over a two-year period to get high quality trees; that is, one graft each year (two grafts). (See drawing.)

The interstem, the rootstock, and cultivar must be hardy for the area planted. North of zone 5, one should use Columbia, Borowinka, Dolgo or other cold-hardy rootstock. The interstem then should be more cold hardy than the Malling series; I suggest P.2 or Ottowa.3.

These interstem trees often have size variability among them and they do not stand alone unless the top 3 feet of the leader is defruited until the trunk and leader are quite strong. I had a Mutsu and a Baldwin on an interstem and I let the leaders fruit. They both started to tip over, and a single-wire trellis had to be installed to straighten them so I could let the trees fruit heavily while young. Once they start to lean, it is difficult to correct without support and/or pruning to stiffen the leader. Most every nursery that sells interstems with a strong rootstock such as MM.111 say they are free-standing. That assumes the tree buyer knows how to prune and manage them. Any young apple tree, with or without an interstem, is safer with a sturdy stake or wire support, at least for a number of years.

Smaller nurseries are more inclined to produce interstems, especially in those areas that depend on rainfall and do not irrigate. Interstems are not of much value here in my area as we have to irrigate anyway, so why not use the appropriate dwarfing rootstock without the interstem?

It seems there are two main reasons for using interstems:

1. Where you have no irrigation and rely on rainfall that often does not come. Then the drought resistance of the MM.111 rootstock becomes important. After three or four years of growth, I build a mound of dirt up around the tree so that half the interstem is covered. Mounding up and getting the rootstock deeper in the ground should eliminate collar rot in the MM.111, as that disease occurs right at and just below soil level. The M.9 rootstock is quite resistant to collar rot, so I bury it half-way (about 4 inches.) This interstem combination produces about the same size tree as M.26.

2. For use in the far north of USDA zones 3 and 4. The cold-hardy seedling rootstocks of Columbia, Borowinka, etc., might give trees larger than desired unless an interstem is put between the rootstock and cultivar to hold tree size down. This interstem would need to be a cold-hardy one such as Ottowa.3 or P.2, both of which have proven cold hardiness.

I cannot test for cold hardiness in our mild climate and have relied on information from personal study in Minnesota, information from growers in cold climates, and from information listed in the back of this book.

<u>M.7A and M.7 EMLA: With spur strains.</u>

The characteristics of this rootstock are described in the table "Rootstocks of Choice for Free-Standing Trees." It is included in this section too because, with spur strains, the trees are about the size of non-spur cultivars on M.26 roots. Remember, spur strains have so many fruit buds that they are dwarfing 20 to 30 percent because much of the tree's energy goes to produce apples. All the other rootstocks in the above trellis chart are not vigorous enough for spur strains unless you would be happy with tiny trees.

Suggested Trellis Spacing Chart
for these size-controlling rootstocks
(measurements in feet)

Size-controlling rootstock	Suggested max. tree height	In-row spacing by tree group size (see chart, next page)				Row width
		A	**B**	**C**	**D**	
Most dwarfing: P.22 M.27	6	4	5	5	6	8 to 10
Dwarf: M.9 P.2 B.9	8 to 10	5	6	6	8	10 to 12
Semi-dwarf: EMLA.9 MARK M.26 Ottowa.3 interstem M.9/ MM.111 root	10 to 12	5	8	10	12	12 to 14
Semi-dwarf to semi-vigorous: Spur strains on M.7A	10 to 12	6	8	10	12	14 to 16

Some variation in maximum tree height may be desired depending on climate, soil, culture, rootstock and pruning. Do not use spur type cultivars on the above roots except M.7A, or the result will be tiny trees with reduced vigor–unless that is what you want. The above tree size groups are applicable only for the above size-controlling rootstock.

☞ Usually plant rows north-south for best light penetration.

Fifty cultivars
grouped by
general vigor and tree size

A. Small

Empire, Freyberg, Haralred, spur strains, Kandil Sinap.

B. Moderate

Arlet, Ashmead, Blushing Golden™, Brock, Braeburn, Coromandel Red, Cortland, Cox Orange, Delicious (standard), Elstar, Erwin Baur, Fuji, Gala, Ginger Gold™, Golden Delicious (non-spur), Grimes Golden, Hudson Golden Gem, Idared, Jonagold, Jonathan, Karmijn de Sonnaville™, Keepsake, Kidd's Orange Red, Liberty, N.Y. #429, Orleans, Roxbury Russet, Spitzenberg, Stayman, Sturmer, Sweet Sixteen, Tydeman's Late Orange, Wickson Crab, Winesap, York.

C. Large and vigorous

Bramley, Calville Blanc d'Hiver, Granny Smith, McIntosh, Melrose, Newtown, Rhode Island Greening, Spigold.

D. Very large and vigorous Gravenstein, Mutsu, Northern Spy.

Trellis Spacing Chart Explanation

The spacing distances suggested in the chart are intended for small trees grown on a three- to five-wire trellis. Distances suggested are satisfactory if you use minimum nitrogen after the third year and trees are grown to the suggested height while fruiting heavily.

Spur strains are usually, but not always, bud mutations of the original cultivar and have many more, closely arranged, fruit bud spurs. This heavy fruiting is about 25 percent dwarfing if compared to non-spurs on the same rootstock.

Bud mutations are thought to be the result of cosmic radiation or, more probably, ultraviolet rays and are also known to come from purposeful irradiation by researchers.

Height of the tree, on the trellis

These guidelines assume good soil, supplemental water (probably spitter, drip or soaker hose) and adequate but minimum nitrogen use. For every 2 feet you *lower* the height from that suggested; e.g., between 12 and 8 feet, you should plant the trees 2 feet wider apart in-row. You can then also move the between-row spacing closer by two feet, because sunlight from these lowered trees (if rows are north-south) shines over these lower trees longer.

Please bear in mind that an extra 4 feet of tree height on a trellis will produce about one-half of a 42-pound-box more apples on trees six or more years old (a full box in some instances).

Perhaps the following true story about a 10-acre orchard will help you to understand this. I started to advise the owners when the orchard was nine years old. All cultivars were grown on M.26 rootstock except Gravenstein, which was on M.9. In their rich soil with supplemental water, all trees had been planted a little too close. All cultivars were planted between 5 feet or 6 feet apart, except Mutsu, which was 7 to 8 feet apart. The Mutsu would easily have filled out an in-row spacing of 12 feet and the others probably should have been in-row spaced 8 feet apart as the tree spacing trellis chart shows.

Because almost all the European Malling rootstocks have been proven to be larger than our nurseries originally suggested, many of the growers 12 to 20 years ago planted trees on rootstock too large for their spacing. On my chart, M.26 is called a semi-dwarf, with two groups listed that produce smaller trees. By the fourth or fifth year, the above-mentioned orchard on the non-spur trees had reached 8 feet in height, where the owner had intended to stop them. His poles were 11 feet above ground, and by putting another wire in at 11 feet he allowed the trees to grow another 4 feet to 12 feet high, just above the top wire. After observing the orchard between the 9th and 12th years, these same trees began producing about one-half 42-pound box more per tree from the now 12-feet-high trees. Since then, by pruning they have been kept at that height, or slightly higher.

The vigor of a tree is first controlled by a certain rootstock, then by the distance between trees in-row and by its end height, or a combination of all three. Without the trellis chart, these factors are often guesswork. In the 10th year, an acre of Golden Delicious in the above-mentioned orchard produced 2,300 42-pound boxes. Sold at an average of $10 a box (a poor price), that acre grossed, before costs, $23,000. So even with somewhat wrong spacing it was still a winner; however, now in its 16th year, most of this orchard, although contained, requires more care to hold it within this spacing. It could be time to saw off every other tree at ground level and let the remaining trees fill in the space sideways, as they should not be allowed to grow higher because of the row width distance. Fortunately the rows had been planted between 14 feet to 15 feet wide; nonetheless, there is barely enough sunlight shining on the bottom 4 feet of branches to keep the bud spurs alive (planted north-south as the orchard is).

The point is that in an orchard with 520 trees per acre, about 250 boxes more per acre were produced without the cost of planting more trees. The solar energy and the additional airspace were free. The additional costs of spraying, pruning and picking of the extra 4 feet were a fraction of the additional income. Everyone was a winner. A home orchardist, of course, may not want trees over 8 feet high and does not need to have them with the correct rootstock-cultivar combination.

In comparison

How about a free-standing orchard with a skilled grower? At about 300 trees per acre, can such an orchard reach 1,000 boxes per acre by the fifth to sixth year on an interstem, or M.7A rootstock? I have seen a number of these orchards. Indeed, my own largest and oldest M.7A rooted cultivar, a Melrose, gives me four to six boxes a year. At 300 trees per acre, one needs 3.3 boxes of saleable apples per tree to reach and hold the minimum of 1,000 boxes per acre that I believe is absolutely necessary for a small commercial orchard to make a reasonable profit. Home orchardists have different goals.

The advantage of growing trees with the roots shown on the trellis chart is that they should return substantially heavier crops and more cash the first five to eight, maybe even 10 years. After that, the bigger trees are as productive even though only half as many, but the smaller trellised trees cost less to spray, prune and pick and usually have better quality apples.

It would only be fair to report that the type of trees returning the most money and world record production, after 10 to 14 years of growth, are probably on free-standing trees in New Zealand. They make use of more vertical air space.

Granny Smith (the original, not the spurs) is one of the highest producing annual apple cultivars in the world. New Zealand growers use the semi-vigorous rootstocks Merton 793 and MM.106, and prune to what looks like a central leader, modified axe system, but makes great use of air space. These trees are taken to 14 or more feet, classified as medium height and rather narrow, and produce 5 to 10 boxes per tree in a mature orchard. Other cultivars such as Braeburn are heavy producers also, although Braeburn is somewhat biennial.

One of the most important factors in this high production in some orchards is a self-propelled machine with a man on a hydraulic boom who has all controls, and can move this small, 4-wheeled machine around the orchard as well as pick the top part of these trees. The lower 6 to 7 feet of the tree are usually picked from the ground. Pruning and thinning are also done by this fine machine. I have seen similar but much more expensive and complicated machines used in our U.S. cherry orchards.

The Merton 793 rootstock mentioned above is one parent of Malling-Merton 111, but it is not produced commercially in the U.S. as its offspring MM.111 seems preferred, although they are much alike. There is now commercial, mature acreage in New Zealand consistently producing 1,600 to 2,000 boxes per acre per year. Of course, they do have a very low-stress climate, summer and winter, and I understand their growers are among the most skilled in the world. I understand from personal conversation that one Granny Smith orchard holds the world record of over 4,000 boxes from an acre.

When I was growing up in southern Minnesota in the 1930s, a stressful climate, we normally used only crab seedling rootstock. Fortunately, our two main cultivars Haralson and Wealthy were genetically dwarfing and we could hold our trees to a maximum height of about 10 to 12 feet. Today, we have many rootstocks for size controlling, cold climates and resistance to certain diseases, so the rootstock sections and the spacing charts are very important for study. Fifty years ago we would have had the whole orchard in rather variable seedling rootstock and almost certainly would have had much lower productivity in the first seven to 10 years.

The day of the big apple tree appears to be over. Europeans deserted these big trees in most countries long before we did. Try these smaller trees and you will probably never turn back. The smaller rootstock trees will need closer attention for the first few years, but in later years, in my experience, they do not present the care problems of the large, vigorous seedling rootstock trees.

Rootstocks of Choice for Free-Standing Trees
From approximately one-half of standard size to standard size
Usually hardy as far north as USDA zone 5
(measurements in feet)

Rootstock	Maximum tree height	In-row spacing (see chart, below)				Row width
		A	**B**	**C**	**D**	
Semi-vigorous						
M.7A	10 - 12	8	12	14	16	16 - 18
MM.106	12 - 14	10	16	18	20	16 - 20
MM.111	12 - 18	12	16	18	22	18 - 22
Seedlings (standard)	18 - 20	Spacing cannot be provided due to genetic variation in seedling rootstock and insufficient test information.				22 - 25

For spur strain cultivars only. Seeds are from current cultivars, primarily Red Delicious, or McIntosh.

---

50 Cultivars Grouped by General Vigor and Tree Size

A. Small Empire, Freyberg, Haralred.

B. Moderate Arlet, Ashmead, Blushing Golden™, Brock, Braeburn, Coromandel Red, Cortland, Cox Orange, Delicious (standard), Elstar, Erwin Baur, Fuji, Gala, Ginger Gold™, Golden Delicious (non-spur), Grimes Golden, Hudson Golden Gem, Idared, Jonagold, Jonathan, Karmijn de Sonnaville™, Keepsake, Kidd's Orange Red, Liberty, N.Y. #429, Orleans, Roxbury Russet, Spitzenberg, Stayman, Sturmer, Sweet Sixteen, Tydeman's Late Orange, Wickson Crab, Winesap, York.

C. Large and vigorous Bramley, Calville Blanc d'Hiver, Granny Smith, McIntosh, Melrose, Newtown, Rhode Island Greening, Spigold.

D. Very large and vigorous Gravenstein, Mutsu, Northern Spy.

Rootstocks of Choice

USDA zones 5-9 and parts of zone 4

<u>M.7A (synonym, French Doucin):</u>

This rootstock is centuries old, having been known in France to La Quintinye (1626-1688), the chief gardener for Louis the XIV at Versailles. M.7 was collected in France and became part of the Malling series as selected by Hatton in England during the World War I period.

About 1960, shortly after virus testing was first developed, M.7 was cleaned of harmful virus and called M.7A. Shortly thereafter, M.7A was subjected to an improved system of treatment which removed the remaining virus. This was done by two research stations. The first was the East Malling-Long Ashton Stations (EMLA) in England, and resulted in EMLA.7. The other was the Washington State University Research station at Prosser, Washington, where Dr. Paul Fridlund developed what is called the IR-2 series. This rootstock was then known as IR-2.7. Only the EMLA and IR-2 rootstocks are known to be free of harmful virus and, as a consequence, may be slightly larger than M.7A. M.7A is thought to have retained two latent viruses that are not exposed except when grafted to a virus indicator cultivar such as Lord Lambourne or a virus-sensitive crab. M.7A is still being used by many nurseries which propagate trees.

M.7A is subject to crown gall when planted in infected soils but has quite a good resistance to *Phytopthera* (collar rot). It is much more resistant to fireblight than M.9, which is important because it sometimes suckers. Most nurseries graft M.7A about 8 to 12 inches above soil level instead of 5 to 7 inches, so that the tree can be planted deeper to reduce suckering and minimize the likelihood of fireblight entering the roots. In order to reduce or eliminate suckering, do not cultivate near the tree and keep voles and gophers from damaging roots.

Of all the *smaller* rootstocks, M.7A appears to work the best in more climates and soils. Trees in North America continue to produce well 35 or more years after planting. Some M.7 rooted trees in Europe are reputed to be over 100 years old. This rootstock is widely adaptable.

In the severe test winter of 1983-84, unmulched trees on M.7A roots died at the University of Minnesota and other locations in zone 4. Rather than M.7A in USDA zone 4, I recommend one of the very cold-hardy rootstocks, which is described in the chapter "Cold-Hardy Rootstocks for Apples."

<u>MM.106, no synonyms:</u>

This rootstock is the result of a combined breeding program undertaken in 1924 by the John Innes Institute of Merton, England, and the East Malling Research Station. The rootstocks bred from this program have been given the name Malling-Merton, now shortened to MM. The primary objectives of the program were to develop resistance to woolly aphid, provide size control, simplify propagation and increase fruit production.

The M.9, M.26, and M.7A rootstocks, which produce smaller trees, are more in demand for the new *non-spur* cultivars such as Gala, Braeburn, Jonagold, Melrose and Fuji. As a result, unless requested on a contract basis, one Oregon fruit tree nursery has ceased producing MM.106 rootstock.

MM.106 is noted for early, heavy cropping and seems to be heavily used by growers in New Zealand and Australia for non-spur cultivars. Granny Smith, for example, crops very heavily on it and seems to retain enough vigor so it does not runt-out, as with some spur strains.

The susceptibility of MM.106 to collar rot on roots in heavy, poorly drained soils and a tendency to defoliate late in the season make it an uncertain rootstock in those soils. It frequently freezes out in parts of zone 5 and just does not survive a test winter in zone 4 of the upper Midwest. It is also susceptible in the eastern U.S. to death from brownline decay, which is a virus (tomato ring spot virus) apparently spread by the dagger nematode, a soil organism.

<u>MM.111, no synonyms:</u>

This is a cross of Northern Spy with Merton 793. Since Merton 793 has Northern Spy as one parent, MM.111 clearly has strong Spy heritage and resistance to woolly aphids. It usually produces a tree about 80 percent the size of standard with non-spur cultivars. Currently, it is used primarily under spur strains of Red Delicious and other spur strains, and as the root for interstem grafts. It is quite resistant to drought because of a large root system. It is also tolerant of both heavy soils and light soils. It is easy to propagate, but it is not an early producer under non-spur cultivars.

<u>Seedling (synonym, Standard):</u>

By this term I mean the rootstocks resulting from planting seeds of various apple cultivars. In the last few hundred years they have been widely used as rootstock for grafting a known and useful cultivar. They can grow to a height of 30 to 40 feet. I suggest that they be used only under *spur* strains of apples, as multiple spurring is about 25 percent dwarfing.

Seeds of Red Delicious, probably pollinated by Golden Delicious and/or Winter Banana, have been found to produce trees with minimal genetic variation. Red Delicious seedlings are also probably hardy about one zone north of Golden Delicious seedlings, so they should be hardy in USDA zones 5 and 6.

In zones 3 and 4 more cold-hardy seedling rootstocks are used, often Antonovka, though if they are pollinated by middle latitude (less hardy) cultivars, the seed's cold hardiness may be diluted.

It must be borne in mind that the cultivar used also contributes to the length of tree life. Many of the Wolf River cultivars grafted to seedling roots in the 1890s are living today, while shorter-lived cultivars such as Spitzenberg are long since aromatic firewood or soil organic material.

One of these old giants on seedling roots is nothing short of awe-inspiring. The Willamette Valley, where I live, has quantities of them around old homestead sites, some with the buildings long gone but with the old trees still producing apples. Most of the cultivars were Northern Spy, Gravenstein, King and Baldwin. One summer I volunteered to help care for a small seedling rooted orchard; most of the trees were from 50 to 80 years old, but I soon decided I did not have enough life insurance, time, or appropriate equipment. These trees are living today, untended, and what fruit they have is wormy, scabby and a source of infection for neighboring orchards. We can grow better apples today on small trees, for which the inexperienced can care without power equipment or tall ladders.

Another issue to consider before planting seedlings is the source of the seed. Seeds can be obtained from canning and juice processing plants. Recently, flowering crab apples have been widely planted as pollinizers in commercial orchards. What effect will this have? Will trees be smaller? Will they be more sensitive to latent virus in various cultivars? Probably. The best response to these issues that I can think of is to use clonal rootstock, except perhaps in zones 3 or 4 where cold-hardy seedling rootstock of Borowinka, Columbia, etc., are not yet diluted with too many unknown genes.

Small home orchards with limited space and small commercial direct-market orchards have a marvelous choice of clonal rootstocks which produce small, manageable trees at lower operative costs than 50 years ago.

Knowing which rootstock to use is the first and most important factor for success in growing quality apples.

Cold-Hardy Rootstocks for Apples
Plus Very Cold-Hardy Cultivars
USDA zones 3 and 4

Cold hardiness in apple trees, especially in the above zones, involves a number of factors. It begins with the cessation of limb growth, usually in August, when the terminal bud on the end of each branch stops growing. Until the tree defoliates it is not very cold-hardy, so the rather rare, sudden early drop of temperature of 30°F to 70°F in October or even early November, while leaves may still be on the trees, can be devastating. Such a severe early storm struck Minnesota, Iowa and adjacent states when the temperature dropped from 60°F to -5°F in 30 hours. This was the infamous Armistice Day storm of November 11, 1940, and it is described in more detail under the history of Delicious.

With each step, the tree goes from cessation of growth, defoliation, early winter hardiness to deep cold hardiness in January and February. During this time a number of physiological and chemical changes take place. Some of these changes are not well understood. Like the human body, these trees have their secrets regarding how they cope with extreme seasonal stress. For serious analysis of cold hardiness, I suggest study of *Temperate Zone Pomology* by Dr. Melvin Westwood, available from Timber Press in Portland, Oregon.

Once the tree breaks out of its rest and opens its flower bloom, it again has little or no hardiness. That is why an area noted for spring frosts of 4° to 8°F below freezing (24° to 28°F) can cause severe flower damage and may result in considerable crop loss. Such a frost pocket area should be avoided if possible. Planning of a small commercial orchard may allow some choice of site; however, home orchardists usually have to plant where they are.

It is interesting to check the maps of countries in the *north* temperate zone and see that perhaps as much as 70 to 80 percent of all apples are grown above the northern 44th parallel. In the north temperate zone of the U.S., one-half of the apple crop is produced in a relatively small geographical area lying between 46 degrees and 49 degrees north in eastern Washington. This area is primarily in USDA zones 5 or 6, and the winter climate is somewhat mild because the Pacific Ocean is not far away, though occasionally temperatures of -20°F do occur in winter. Because the eastern Washington climate is usually relatively mild, not as cold as the Midwest, the Malling rootstocks have been successful and the really cold-hardy rootstocks are usually not used.

The 44 N parallel runs through southern France and northern Italy, so much of northern Europe's, the British Isles' and Scandinavia's production is produced in these northern latitudes. The effect of large bodies of water allows such plantings. There are sufficient frost-free days to mature a great many cultivars, and the winters near these large bodies of water are usually not severe enough to harm the trees or rootstocks. As we get into the interiors of Poland, East Germany and European North Russia, of course, the situation changes and hardier rootstocks and cultivars are required. Even Finland grows apples commercially, though near the water at 60 degrees N latitude, using cold-hardy rootstock and cultivars.

The southern hemisphere is beginning to grow some quantities of apples, and most come from southern latitudes between 30 degrees and 50 degrees south latitude. New Zealand, Tasmania, Australia, Argentina, Chile, Peru and South Africa are in these southern latitudes and produce fine apples, but not yet in the same quantities as northern temperate zones, and many use Malling rootstocks.

Where cold-hardy rootstocks are needed (USDA zones 3 and 4)

The oceans are too far away to soften the harshness of winter and summer in central Canada and the north-central U.S. There are fewer frost-free days, and cultivars that ripen earlier, in August or September, are safer to grow. It is not historically successful to attempt growing considerable acreage in USDA zones 3 or 4 commercially.

Minnesota commercially produces only about 500,000 42-pound boxes a year, mostly in the south and southeastern section (zone 4), from about 100 small commercial orchards. They could produce much more. I lived in the Northfield, Minnesota area for a third of my life, and growing apples *was* historically successful on our farms. Most farmers had Wealthy, Duchess of Oldenburg, Haralson, Whitney Crab and other cold-hardy crabs such as Hyslop, Dolgo and Martha. In contrast, today there are 30 cultivars grown (or under test) in Minnesota, but not many on the farms anymore. The first testings of record were done by a famous amateur apple breeder named Peter Gideon about 1850. His efforts in Excelsior, Minnesota, were continued by the University of Minnesota, and since this early start, about 15 above-average cultivars have been bred, starting with Wealthy in 1860 by Gideon. Sweet Sixteen, Honeycrisp and Keepsake are three recent results; all are above average apples.

More apples could be grown in these areas commercially, and I believe they will be because more attention is being paid to the truly cold-hardy rootstocks. Growers should now know that M.26 should not be used as a rootstock without supplemental summer water available, although it has some cold hardiness. It is very drought-susceptible and has failed from coast to coast where grown in certain states without irrigation. It is one of the favorite rootstocks here in the west for trellised orchards because irrigation is used. Most of these western U.S. areas are not highly susceptible to fireblight or collar rot, or M.26 would probably not be used as much, as it is susceptible to both, especially fireblight.

The size-controlling rootstocks were discussed in a previous chapter, but another set of rootstocks listed below are much hardier and are grown largely from seed and not cloned vegetatively, although there are some clones of Antonovka reported by researchers. You should be aware these seedling plants were pollinized by something, so there will be some genetic variation in them.

Those mentioned below have been regarded as near standard in size; however, some are more the size of 60 to 80 percent of standard. Part of this may be because they are used in zones of high stress, both summer and winter (zones 3 and 4), with shorter summer seasons and long, dehydrating winters. Genetics also plays a part.

The Canadians seem to be doing the most research on these rootstocks as they have the most need for them; however, U.S. growers in the upper midwest and New England also need them. All are in the *Malus* genus. These listed below are available in Canada (some in the U.S.).

• Columbia Crab: One hardy rootstock used by Bailey Nurseries of St. Paul, Minnesota. Compatible with most cultivars. For zones 3 and 4. Recommended, especially in high pH soils (7.5 to 8.5 pH).

• Borowinka (synonym, Duchess of Oldenburg): Known better in Europe. Very compatible. For zones 3 and 4. Available in the U.S. Used extensively by Lawyer Nursery of Montana. Highly recommended.

- Renetka: Another European rootstock, hardier than Antonovka. For zones 3 and 4. Rather rare in the U.S. except from Lawyer Nursery.

- Anis: Reported to be hardier than Antonovka in Canada. Widely used in northern Europe and offered by some nurseries in the U.S.

- Antonovka: Easy to propagate. Compatible. Usually grown from seed but there are now clones, some of which appear resistant to collar rot. Can induce low production and some blind wood (no fruit buds). Widely available.

- Prunifolia (synonym, Plum Leafed Apple): The most extensive and fibrous roots of apple rootstocks. More hardy than Antonovka. For zones 3 and 4. Available from Lawyer Nursery as a rootstock.

Malus baccata (Siberian crab) and *Malus bittenfelder* are also cold-hardy rootstocks, but have incompatibility with many of the U.S. apple cultivars and probably should not be used.

Hardy Apple Cultivars Suggested for Cold-Hardy Rootstock
USDA zones 3 and 4

1. Carroll	9. Honeycrisp	17. Redwell
2. Cortland	10. Keepsake	18. Red Astrakan
3. Duchess of Oldenburg	11. Lubsk Queen	19. Sandow
4. Fameuse	12. Mantet	20. State Fair
5. Garland	13. Mandan	21. Sweet Sixteen
6. Goodland	14. Melba	22. Wealthy
7. Haralred	15. Norland	23. Westland
8. Hibernal	16. Patterson	24. Wolf River

Sweet Sixteen, Keepsake, Haralred, Honeycrisp, and Cortland are in the descriptions. I did not include the others above in the descriptions because of lesser use, primarily for the cold zones, and also because of a lack of statistics.

Hardy Multi-Purpose Crab Apples

1. Centennial	6. Rescue
2. Chestnut	7. Whitney
3. Dolgo	8. Wickson
4. Hyslop	9. Young
5. Martha	America

Centennial, Young America, Wickson, Whitney and Martha are very good for fresh eating. All are good for pickling and in cider mixes. Dolgo makes fine jelly, and some others do too. Most have large, beautiful blossoms.

Planting Guide

Number of Trees or Plants Per Acre
43,560 sq. ft. = One Acre

Space between rows of trees (in feet)

Space between trees in the row (in feet)

	6	7	8	9	10	12	14	16	18	20
4	1815	1627	1361	1210	1089	907	777	680	605	544
5	1452	1240	1089	966	871	726	622	544	484	435
6	1218	1037	907	806	726	605	518	453	403	363
8	907	777	680	605	544	453	368	339	320	272
10	726	622	544	484	435	362	311	272	242	218
12	605	518	453	403	362	302	259	226	201	181
14	518	444	388	345	311	259	222	194	172	155
16	453	388	339	325	272	226	194	169	151	136
18	403	345	302	268	242	201	172	151	134	121
20	363	311	272	242	218	181	155	136	121	108

Cornell-Geneva apple rootstocks

This is primarily a progress report, as the Geneva Experiment Station at Cornell University in Geneva, New York, is just releasing these rootstocks to selected nurseries for propagation, and these rootstocks and/or cultivars grafted to them will not be available until about 1996 or later.

This information is an update taken from the oral presentation given by Dr. James Cummins, research horticulturist at the Geneva Station. These rootstocks will be tested at the NC-140 test sites all over North America, but these test sites do not sell rootstocks. Selected nurseries sell them.

The European rootstocks (primarily the Malling Series) have certain difficulties in parts of North America due to our different soils, pathogens, insects and climate. Most of these new rootstocks from the New York breeding programs are claimed to be more resistant to fireblight, union necrosis, *Phytopthera*, woolly apple aphids, replant disease, higher summer soil temperatures and sensitivity to drought. Also, they are economically feasible to propagate, and are claimed to be tolerant of the latent viruses contained in cultivars commonly grafted onto them.

Out of over 325,000 seedling hybrids of known lineage, 15 "elites" will be licensed for clonal reproduction to selected nurseries. These are rootstocks that will vary in tree size similarly to the Malling Series from M.27 to MM.111.

Meadow Lake Nursery in McMinnville, Oregon, will propagate four of these rootstocks. The Geneva Station is also licensing other nurseries, unknown to me as of this writing.

I believe the Cornell-Geneva apple rootstocks should be mentioned, as they may very well have the same impact on apple growing in the 21st century as the Malling Series has had for the last 50 years. The facts will come from testing.

Part V

Physiological Apple Problems

Netting (induced russeting)

Certain apple cultivars are susceptible to this condition. There seem to be three main causes:

1. Certain early sprays, especially early sprays for apple scab fungus, often cause netting. Controlling scab seems more important. Golden Delicious is one of the most susceptible.

2. Powdery mildew infection can, in turn, cause the physiological problem of netting. See "Powdery mildew" under "Fungal Apple Diseases."

3. Netting can occur with frost just after bloom and/or cool drizzly spring weather from the time blossoms drop until apples have distinctly formed (three to five weeks). This problem could also be related to (1) and (2) above, and it is often unclear which is the real culprit. Poor weather alone seems to increase the netting in certain semi-russeted types such as Ashmead, and Kidd's Orange Red. This is a fuzzy area and my observations have led me to fatalistically declare that we can't do much about the weather anyhow.

Genetic susceptibility is surely a factor too. After all, growing apples can be likened to rearing children in that we are never absolutely sure of the outcome.

Bitterpit

I think this is the most important and potentially devastating of the physiological problems. These corky spots in apples appear to be primarily caused by a shortage of calcium in the apple and not necessarily in the tree or soil. Although it is possible all three could be deficient, that is unlikely.

Most primitive apples that grow wild over large areas in Asia are small in size and seem to receive enough calcium through the stem. Our modern, heavily cross-bred apple cultivars occasionally spawn very large apples, and the physiological potential to transfer calcium into a large apple is often inadequate under certain weather and moisture conditions. Since the smaller-fruited apples such as Jonathan, Lustre Elstar and Gala never show bitterpit in my orchard, it seems logical that much of the calcium must go into the apple during its formative stages and is sufficient up to a certain size. Large apples, such as Northern Spy, Jonagold, Spigold and Mutsu, are usually more susceptible. Younger trees are *much* more susceptible to the problem. Let us examine what seem to be the primary causes of bitterpit and how to deter it.

These causes are my perception, in my climate and from my tests in my test orchard or in small commercial orchards nearby. They may not agree completely with certain research in other climates, and need not. Consult with your extension agent for specific information for your area.

- Too much growth in tree branches (after the third year) caused by excessive nitrogen. In clay or clay loam soils there is usually sufficient calcium in the soil. New growth of over 8" to 15" on bearing trees means much of the calcium is used in these new branches and leaves, and perhaps not enough is going into the apples. That is the problem, getting the calcium into the larger apples.

- Insufficient water from late June through early September. If the tree does not get sufficient water after transpiration through the leaves, then it may pull some liquids from the apple, and minerals such as calcium will come out too. Sufficient summer water then is paramount, especially on M.26 and MARK rootstocks which are quite drought-susceptible.

- The pH (negative log of potential hydrogen ion concentration) is too low; a pH of 5.0 to 5.4 means calcium is probably not being maximally absorbed by the tree. One scientist speaking to a group of home orchardists said that adding lime does not help contain bitterpit. That is misleading! Proper pH (6.5 to 6.8) helps contain bitterpit and lime may need to be added to bring the pH up. True, there are other factors involved. Some observers disagree with me on this point, but then sometimes bitterpit causes are difficult to pin down. Extreme heat (95º to 110ºF) for sustained periods makes bitterpit very difficult to control. Perhaps certain apples are better not grown in such a climate.

- There probably is some genetic tendency in certain cultivars, particularly in certain climates (e.g., cork spot in York, and in Braeburn in hot Southern California). Consider Newtown in the Willamette Valley climate, in contrast to the Hood River Valley in the Parkdale area of Oregon, which is at 2,000 feet and consequently cooler. If watered properly all summer and picked in late October or early November, we do not see bitterpit in Newtown here. Hood River, in a cooler climate, does see some bitterpit in Newtown. Also I believe too much nitrogen and picking three to four weeks early, which some packers demand, is poor management and can lead to this problem.

Hot areas such as the Yakima Valley in Washington seem to have far more bitterpit problems in most cultivars, except Gala, than we do. They have sufficient water, so it appears to be related more to excess nitrogen and, in some years, too many 90º to 105ºF days, and from early picking for long storage.

Some years ago my *first* crop of Northern Spy on M.7 rootstock had bitterpit in about one-third of the apples. The tree was on a high, dry mound, had very little summer water, was a young tree and had three feet of new growth as a result of too much nitrogen. I filled in around the mound, mulched it heavily, watered it thoroughly all the next two summers (it had no apples in the fifth year), and in its sixth year it had no bitterpit. This was accomplished by better management on my part, and that tree did not show any bitterpit symptoms before removal a few years later to make room for testing another cultivar. One can say it was luck or the weather, but the least we can do is remove factors that we know contribute to the cause of bitterpit.

Other ways of preventing bitterpit:

- Try to not let apples get too large. For example, Jonagold is such a fine near-annual bearer in my orchard, you might let it crop rather heavily so apples do not become too large.

- Watch the levels of boron and zinc. Two orchards nearby and only a mile apart displayed virtually no boron or zinc (micronutrients) in soil and leaf analysis and developed serious lumpiness and bitterpit in Jonagold. This was corrected with foliar sprays the next spring.

• I think what really can save the crop in susceptible cultivars and less than perfect climates is calcium chloride sprays. They must touch and cover the fruit. These sprays are low cost. I believe it is important to spray all through the growing period to within two weeks of harvest. A fine Jonagold grower in the mid-Willamette Valley now sprays calcium chloride six to eight times a summer and keeps apple size at medium to medium large. I have seen no bitterpit in his Jonagolds now for three years. Braeburn also receives many calcium chloride sprays here in our valley. The New Zealand orchards spray many times per season, according to fellow researchers who have been there.

One cannot violate good management programs and expect calcium chloride to completely eliminate bitterpit, but I cannot overemphasize how helpful these sprays are. Correct amounts are important to prevent leaf burning. Your extension agent can give you the proper amounts to use.

Remember, calcium is critical in cell structure, and without enough of it an apple will show you its need via lack of firmness, storage breakdown and even bitterpit.

In my cooler climate, other physiological problems are relatively unimportant to non-existent. The Portland, Oregon, U.S. Weather Bureau says the average number of days over 90°F from May 15 to October 15, is only 10.7 days. Low nighttime temperatures from about August 25 to November 5 (seldom freezing) contribute to an ideal, rather dry summer climate for almost any apple cultivar from Elstar to the very long season Sturmer or Pink Lady. Our three main problems in the valley are apple scab fungus, powdery mildew fungus and codling moth insect (see under "Insects" and "Fungal Apple Diseases.")

Apple Orchard Culture Management:
Pruning, disease, insects, cold country culture management (zones 3 and 4) and soil

Pruning

People constantly inquire as to why they cannot find a step-by-step, snip-by-snip book that allows them to prune all sizes of apple trees in a scientific and exact manner. A new book *Training and Pruning Apple and Pear Trees*, by C.G. Forshey, D.C. Elfving, and R.L. Stebbins, certainly does help to supply guidelines, but it gets well into plant physiology. It gives rather accurate pruning tips for specific tree forms. Some of the more intricate shaped systems, such as the flat, horizontal Lincoln Canopy, are based on a future potential need for mechanical harvesting of apples. These are complicated, expensive systems that after many years of trial in New Zealand, Australia and our research stations, do not seem cost effective to me. Such systems will not be described, as home orchardists and small direct market growers may have little interest in them. Many readers, though, will have an interest in a simple three- and four-wire trellis system, as this solves many problems and does not require ten years of training or a doctor's degree in plant physiology.

We should have general sound guidelines and then not worry too much about the results, unless we find we should correct something for a particular cultivar. Apple trees are most forgiving and seldom object to how they are shaped.

Why are there few, if any, simple books on pruning? Because on the same rootstock, ten trees of the same cultivar down a row will all be pruned somewhat differently because the branches are not all in the same place on each tree. The general shape you want in the tree is then the guideline. Pruning is more common sense and an art form than a science, and is learned primarily by doing. Forget any fears of mistakes, follow the guideline and use reason. Determining the appropriate tree shape is based on the need for sunlight on the fruit buds as they will die without sun. Sunlight is also needed on the apples, especially red ones, so they color well and get the proper levels of sugar (carbohydrate solids). Green apples such as Newtown and Granny Smith need more shade or they will sunburn, but one can still follow the general guideline.

Initially, most people are apprehensive about pruning apple trees and labor in a fear that is really unnecessary. I know, I went through it and was greatly relieved to find my trees did very well. It seems about the only sure way you can kill the tree *by pruning* is at ground level with a chain saw.

So, now it boils down pretty much to objectives: why are you making that specific cut? If you do not have an objective (whether it be a right or wrong one), then I don't think you should cut it at all, especially the first three to five years.

We are in the age of the small tree due to dwarfing rootstocks, but some growers haven't tried them yet and don't know what they are missing. Home orchardists especially should not be planting apple trees whose size does not allow caring for them easily and safely. Let us list some problems with standard or near standard trees on rootstocks that help make them large, and then you decide whether to grow these huge trees.

- They do not fruit early as a rule, so you wait too long for apples. Northern Spy, Gravenstein, Red Delicious, Baldwin and Newtown, may not fruit for 6 to 12 years on seedling (standard) roots. Size-controlling rootstock, as shown on the Trellis Chart, fruits in 2 to 4 years.

- The large, vigorous tree produces a great deal of growth (wood) and requires considerable pruning, often yearly, to allow light into the canopy. Unless they are overfed with nitrogen, the smaller trees need very little pruning.

- In time, these trees will be large enough to require tall ladders and powerful spray equipment. They will be difficult to prune and will take up too much room. You may need a strong life insurance program to go with the additional hazards, and they are real.

- In city lots or small rural homesteads, large standard trees are almost always untended and unsprayed, and they are a constant source of scab and codling moth dispersion which nearby neighbors may not appreciate.

- You also may not want fifteen or so bushels of one cultivar that each big tree will eventually give. You may prefer one to two bushels each of a number of apple types.

With smaller trees you can do much, or all, of the work from the ground and eliminate 85 percent of the pruning. This is reason enough, and there are other reasons. Dwarfing rootstock can produce larger apples, and on M.9 rootstock will have less biennial bearing.

Pruning to a central leader

Most pruning instructions are based on the central leader form that keeps the top half of the tree from shading fruit buds or apples on the bottom half. A strict central leader tree, whether free-standing or on a wire trellis hedge-like system, is then shaped like a Christmas tree.

A central leader is the top or central branch emanating from the trunk and should be dominant and kept growing straight up until the end height is reached. Then it is cut off yearly, or bent over to slow or stop tree growth.

The French Vertical-Axis system, which is quite new, is a narrow angular central leader system on which the thicker, older branches are entirely pruned out in favor of shorter, newer branches with new buds, starting in about the fifth year. Little or no pruning is done before then. It was designed primarily to lower pruning costs and is quite productive.

Slender spindle

This appears to be a smaller version of the French Vertical-Axis system. Both systems are highly productive if in single rows. They expose the fruit constantly to sunlight and are used in northern Europe (The Netherlands, Belgium, northern France, England, etc.) where they do not get many 85° to 90°F days, and do not worry much about sunburn. In eastern Washington and the hot Milton-Freewater area in northeastern Oregon, this system, unless modified to produce more shade and to use overhead cooling sprinklers, etc., would fry the fruit on cultivars such as Jonagold, Newtown, Granny Smith, Fuji, and other cultivars susceptible to sunburn.

After much trial and error, I have devised a culturing procedure which reduces pruning to the very minimum and is designed for a balance between fruiting and tree growth. We should not be growing firewood, but rather high quality apples, and smaller trees fit such a philosophy. Smaller trees also do not have the added costs of all types of pruning instruments or too much fertilizer.

Nor am I in favor of some of the extremely high density systems for home or small commercial orchards such as triple rows. These rows shade out the fruit in a few years. There is so much poor-colored fruit and so many unsprayed areas in these systems that the amount of saleable apples is probably not more than you get from a simple three- to four-wire trellis on single rows systems. The additional costs of trees and posts per tree, and/or trellis wires for 1,100 to 1,500 trees per acre is usually not justified in net profit. A former colleague of mine used to say he was not interested in most profit figures, he was interested only in the "net-net-net" profit. I am referring to the small direct market grower here, and not the home orchardist. The ideal quantity of trees per acre varies with rootstock; however, I believe 600 to 700 trees to the acre (on the smaller rootstock) is about the maximum that should be planted in order to produce high tonnage per acre, high grade apples and lower planting costs.

In my test orchard my usual pruning tools are only two: the remarkably dependable by-pass Felco pruner and a small, very sharp bow saw. At a garage sale once I bought for $3 a third tool, one of the long-handled (7-foot) pull-down pruners, as occasionally one of my trees gets a 3-foot growth on the top and I can then cut it off from the ground without climbing up a ladder. Loppers are helpful, but I don't like them as well as a saw for cutting a branch close to the trunk, as they often leave a stub too long to heal over and rot sets in the stub.

A program to greatly reduce pruning

Most of this pruning is done around August 15 to let light onto apples that may be too shaded, and after branch terminal buds have formed so you don't get tender regrowth that could freeze.

- Fertilize (top of ground) the first two years to develop these dwarfing trees to 6 to 7 feet by the third year. Then, in good clay loam soils, do not fertilize until the tree shows a visual need, which could be many years later. A visual need is new growth of only 4 to 6 inches or so and leaves turning a lighter color. I prefer nitrogen in the form of calcium nitrate, as it does not change pH as a rule. Other forms such as blood meal, compost and liquid fish emulsion are fine too in small amounts. I use fish emulsion because of its good trace mineral content. Stay away from nitrogen coming from ammonium sulfate, as it lowers pH. Only in soils with a pH of 7.5 or more should ammonium sulfate be used for nitrogen. An exception to the above rule against fertilizing after the second year, is the addition of lime. Add lime each year until the pH is between 6.4 and 6.7, which is ideal for apples. Lime (calcium carbonate) should be turned over with a shovel, as calcium travels downward only about 1 inch per year and does not travel sideways at all.

- Starting in the second and third years, tie down branches so that they either extend out horizontally or upwards at an angle of no more than 45 degrees. This helps induce early fruiting and shut down excessive growth. When the leader has reached its end height (7 to 10 feet on a three- to four-wire trellis), *bend* the leader over the top wire to slow the hormone flow downward that would otherwise make the tree grow upward much faster.

- Let the tree fruit as heavily as that cultivar can without becoming a strong biennial bearer. Observation and practice are called for here. Low or no additional nitrogen is especially important to cultivars that can exhibit very large apples; i.e., Jonagold, Mutsu, Spigold, etc. These apples must not be overthinned as they are susceptible to bitterpit, sometimes highly susceptible. I suggest four to eight calcium chloride sprays for cultivars susceptible to bitterpit, each two to three weeks apart, starting two weeks after blossom drop. The spray must contact the apples to be effective.

- Take at least one soil analysis during the first five years and check for shortages of boron and zinc, as they often leach out in our climate and a severe shortage can make fruit unusable. Be careful with boron; too much is toxic. Also check soil pH and, if it is between 6.4 to 6.7, leaf analysis is probably not necessary. (Leaf analysis costs about $27.) There can be a difference between soil and leaf analysis for a number of reasons. A pH in the low 5s or above 8 is usually the cause. These pH extremes can mean certain nutrients are unavailable to tree roots.

If these points are followed, very little pruning needs to be done. One of my nine-year-old Jonagolds on M.26 rootstock is only 8 feet high and is in a productive balance of annual production, while most branches (80 percent) grow only 6 to 12 inches per year. Only one large and a few smaller branches have been cut out over this nine-year period.

One grower told me he had used no nitrogen since the first year, yet his four-year-old trees had 3 to 4 feet of new growth. Further questioning revealed this four-acre area had been an alfalfa field for five years. He did not know that alfalfa roots produce bacterial nodules that are high in nitrogen and leave residual nitrogen in the soil for years. His Jonagolds had poor color and bitterpit. Recent former usage of your apple orchard may determine soil nitrogen levels, and even though you may not have added nitrogen, if growth is over 2 feet a year, it would appear someone (perhaps a former owner) or some plant did.

Let us avoid the often yearly ritual of dumping some amount of 10-10-10 fertilizer around an apple tree so it will grow large, and then every February or March hauling out six or seven expensive pruning tools and some extension ladders, and climbing precariously around cutting off the extra growth which we need not have created in the first place.

Have we been going about it backward in the past by teaching yearly fertilizing and pruning first? Instead, shouldn't we teach the lower-cost, less work-intensive, proper management program first? Then use the smaller amounts of fertilizing and pruning if needed?

I like clay loam or drained clay soil that is not too rich, otherwise one can get too much yearly new growth with the calcium going into long new branches and leaves instead of fruit. I don't want just an apple. I want apples that have the necessary calcium in the cells for firmness, flavor, and maximum storage. We could call this fine-tuning a tree, and it is very satisfying to observe the trees' positive reactions, with high quality fruit and slow growth that needs very little pruning.

Fungal apple diseases

Apple scab (*Venturia inaequalis*)

Potentially, diseases are many, but if apple scab fungus is controlled early, other diseases will collectively pale in significance, at least here in the west. Orchard sanitation, raking up infected leaves in the fall and burning them, goes a long way to controlling scab. The fungus spores overwinter on these leaves and any infected apples on the ground. Unlike pear scab (*Venturia pirina*), apple scab fungus spores do not overwinter in the tree, so it is a waste of time to spray the apple tree in winter for apple scab. Please do not misunderstand! It is desirable to winter and spring dormant spray your apple tree with copper or lime sulphur or oil for other reasons, but not for apple scab. Fortunately apple and pear scab do not cross-infect one another. Anthracnose fungus, an organism distinct from apple scab fungus, is primarily controlled by dormant, or delayed dormant, copper sprays. Lime sulphur is also helpful.

It is important to spray with a fungicide in very scab-prone areas such as mine, starting at or just after bud break. The weather will determine how many times to spray. I usually spray every two weeks or 10 days from bud break until the rains stop in May or June (three to six times).

The important factor is to keep your orchard sprayed in the *early* stages so it does not go into a rampant secondary spore infection, which comes from the tree leaves instead of old leaves on the ground. If that happens, this otherwise rather easily controlled fungus now becomes a King Kong monster. It need not happen. Timing and habit are important.

Organic growers in climates like mine must make sound decisions. Usually, organic certification allows spraying with sulphur. Of course sulphur helps in low scab areas, but here it seems to be practically worthless. The organic grower should be using those cultivars that have scab immunity, such as Goldrush, Liberty and Bramley.

If you want to grow the high-quality but scab-prone cultivars without using controlling fungicides, then eastern Washington and certain other dry spring areas in the west are the places to live. The organic certification program has very narrow guidelines; consequently certain organic programs do not seem to work in many climates. These certification programs have no money for research, so it is for scientists to solve this very difficult problem. Unchecked, apple scab fungus

253

can bankrupt a commercial grower; however, I do not find it that difficult to stop with a few timed early season sprays. It must be *prevented*. Controls for apple scab fungus usually control other fungus problems. If you need additional and specific information, you might want to consult your local extension agent.

Apples described in this book that are immune or resistant to apple scab fungus

Around the world there are many old-time apples with great scab resistance, but they are seldom being used for breeding—perhaps because many are not bright red and commercial growers and packers are hooked on red apples. Fortunately, the public does not pay as much attention to color as long as the apple is crisp and has fine flavor. If this were not true, then Fuji would never have made it, as it is often a yellowish brown, muddy color. One wag told me he wanted to put up a sign over his Fuji that said "the great tasting ugly apple." This might work!

Below are 13 apples I have described that require little or no spraying for the Willamette Valley strain of apple scab fungus.

Scab immune
Liberty
Goldrush
Bramley

Partially scab resistant in order listed
(During some wet years in April and May these require some spraying)

Hudson Golden
Gem
Roxbury Russet
Erwin Baur
Tydeman's
Late Orange
Ashmead
Orleans
Keepsake
Sweet Sixteen
Jonathan
Cortland

Powdery mildew (*Podosphaera leucotricha*)

This is usually the only other major fungus affecting leaves and apples that must be controlled in orchards in the west. Some of the fungicides that control apple scab do not control mildew. The fungicide directions will explain its uses. Your local extension agent is the person for legal and expert aid in what to use and when.

There is a manual control which helps, since the fungus attacks primarily the apple branch tips. Prune off the end of the branches exhibiting the white cottony leafs, collect in a paper bag, and burn them thoroughly. Prune a few inches below the infection.

Spraying with a mildewcide from tight bloom cluster to petal fall is essential to success in controlling powdery mildew. Unchecked, powdery mildew can severely russet apples.

Cultivars listed in this book that are the most susceptible to powdery mildew are Cortland, Jonathan, Idared and Calville Blanc d'Hiver, although others get it to a degree.

Collar rot, sometimes called crown rot (*Phytopthera cactorum, P. cambivora* and other species)

This fungal disease produces cankers at or just below the soil line in the root-crown area and is usually only a problem of the roots. In the highly susceptible cultivar Grimes Golden, it may climb from the root area up the tree. Because M.9 is quite resistant to collar rot, my Grimes Golden is grafted onto M.9, and I have had no problem.

M.9 rootstock is the most resistant, while M.7 is seldom infected. In low wet areas, MM.106 and M.26 rootstocks are often killed by this disease.

Planting trees on a mound of bermed-up soil can eliminate much of the problem, but no apple trees should be planted in low, saturated areas. In those areas, tiling should be done, but soil structure in clay soils is often not changed for a number of years by tiling, so do not expect miracles. Prevention is the road to success against this disease.

Bacterial diseases

Fireblight (*Erwimia amylovora*)

Fireblight is a very serious bacterial disease of pears; however, in certain spring weather it can also attack susceptible cultivars of apples such as Jonathan, Idared, Rhode Island Greening and others.

Red Delicious, Golden Delicious, Stayman and McIntosh have some resistance and, even if they get infected, the disease seldom enters the main branches or kills the tree.

This bacteria is transmitted primarily by bees, aphids and other insects and by pruning tools. After making a cut, pruners should be disinfected for at least 60 seconds; a Clorox solution or rubbing alcohol could be used.

Usually, no one control will stop fireblight. Prevention is the main control, which is done by spraying at blossom time and shortly after with streptomycin and other antibiotics, which go by various trade names.

Pruning of infected parts at least 5 inches below the visual browning of leaves may help. Burn the prunings.

Hold down new growth by reducing nitrogen use, as the bacteria like new succulent growth.

Only a limited amount of fireblight is seen here in the valley, about every 20 to 25 years when weather conditions favor it. I have never seen it in my work with apple growers in our valley.

Blossom blast (*Pseudomonas syringae*)

This is a bacterial disease that can be very harmful to stone fruits, especially cherry, but that sometimes also affects apple blossoms. It does not ooze in cankers like fireblight on apples. In stone fruits, it kills whole limbs.

Fireblight likes warm spring weather for infection but *Pseudomonas* is a strange critter that likes to ride in on the cold, wet winds of February and March. It is called false fireblight here as it resembles it, but it is seldom serious on apples. It often badly infects pears. Spraying at bud break usually prevents it in apples. Copper sulphate during fall and delayed spring dormancy usually prevents it.

Insects

Codling moth (*Laspeyresia pomonella*), *Olathrutidae* family

Codling moth is thought to have originated in Asia Minor, but it has been a destroyer of apples in North America since about 1800. Unfortunately, it has alternate hosts such as large fruited hawthorns, pears, quince and occasionally cherries.

This small, obnoxious, grayish moth has few enemies because it is a night traveler, starting at dusk, and most insects that might prey on it don't see the female while she lays her eggs. In the daytime they are highly camouflaged. The common apple worm comes from this moth.

Finding out when the moths arrive is a major key to their control. The arrival of the moths varies from year to year by 10 days or more due to weather conditions.

I save cottage cheese cartons, punch a hole on opposite sides of the top portion with a leather punch, then place a looped string in the holes. I fill these cartons about two-thirds full with two parts molasses to seven parts of water, with a couple of teaspoons of vinegar thrown in. I then hang the cartons about six feet high on the perimeter of the orchard. I check them in the mornings for the small gray moth with a vague bronze band across the wings and wing tips. Once you have started catching them, it is time to start spraying. Using enough of these catching cups will no doubt lower codling month populations, but in tests in my orchard, surrounded by many host plants, they are insufficient and some spraying needs to be done.

Organic certification usually allows certain sprays that are derived from plants such as ryannia, pyrethrum and rotenone to be used. Additionally, the certification usually allows use of the granulosis virus (which affects only moths) and the easy to commercially produce bacteria, *Bacillus thuringiensis*. This is sold under trade names such as Dipel and Thuricide. One organic grower in Canada is using all of the above in a constant spraying program, as none of them seem to last long in sunlight. Costs were much higher than using a few chemical sprays per summer but this organic program did work well and, of course, had a low toxicity rating.

One organic grower I met uses pyrethrum almost exclusively and does not understand that it is known, as are most organic sprays, as a wide-spectrum killer, and will kill the "good guys" along with the "bad guys." For example, about one-third of the aphid species eat the other two-thirds. When the "good" ones are killed off, sometimes there are enough "bad" ones around to explode in numbers due to a lack of natural enemies.

Studies all over the world show certain insects are now becoming quite resistant to the constant use of a particular organophosphate (Guthion). I use primarily an organophosphate with the trade name of "Imidan" which is much more selective than pyrethrum, controlling the codling moths without killing many other beneficial insects. I don't believe four to five sprays a summer (from about June 1 to September 15) adversely affect the environment as much as 12 or 17 sprays of pyrethrum. Although the organophosphates biodegrade fairly soon in sunlight, *never* take them lightly at application time. I always wear a hat, a long sleeve shirt and protective glasses. Right after applying, even though I do it in a no wind condition, all my clothes go in the washing machine. Special care must be taken before the chemical is diluted and it should always be locked up away from children *and* adults. Fortunately, Parathion, the most dangerous of this group, cannot be used any longer.

We now have little strips to put in orchards which are infused with the female codling moth hormone. When hung all over the orchard, the hormone so confuses the males that they are often unable to find many mates. This does not seem to work well in an orchard under a few acres. When you have just a few trees you may pull in many times more males than usual, enough so mating does occur. These strips are very helpful, maybe the best control of all, in orchards of 3 or 4 acres and larger, especially if there are no other host plants nearby.

Codling moths and the resultant apple worms seem to have no beneficial reason for existing and if completely ignored in high infestation areas without sprays (organic or otherwise) one might believe it is a tool of the devil to frustrate apple growers. (See picture of the worm on an apple, from codling moth egg).

Apple maggot fly (*Rhagoletis pomonella*), *Tephritidae* family

The apple maggot is a major pest of apples in the eastern U.S. About 10 years ago apple maggot fly was detected a few miles from my home, in southwest Portland. This was brought to the attention of scientists at Oregon State University.

Some of the University's entomologists had come from the eastern U.S. and knew that this fly moves very slowly from place to place. They suggested testing for the fly all the way south to the California border in the next year. They found it: an infestation so wide it could not be expected to be eradicated. This means the fly probably arrived in western Oregon at least 20 or 30 years ago. It remained undetected because commercial orchardists were controlling it by spraying for the codling moth and consequently did not see it. Home orchardists saw a worm in their apple and thought it was codling moth.

Actually the damage caused by the apple maggot is quite different from that caused by the codling moth. The apple maggot tunnels all through the apple flesh, hence the common name railroad worm, whereas the codling moth eats primarily in the core area and exudes a reddish brown sticky frass on the entrance or exit hole. The codling moth worm is much larger and its head is a dark brownish-black while the apple maggot worm's is not.

Since the apple maggot fly starts laying its eggs about one month after the codling moth, sprays used for coddling moth usually control the fly too, except for the eggs that the fly lays in mid-September or later here, after the codling moth has stopped. Some areas in the upper south and southeastern U.S. have reported apple maggot emergence and egg laying three times in a season, so don't ignore that late spray. In 1987 I had not yet detected the apple maggot fly, and that year I did not spray for coddling moth after August 25th—a big mistake. September and much of October were unusually warm and dry, instead of cool with the first fall rains. Apple maggots struck and so did one more hatch of codling moths. For the first time, about half of my apple crop was lost. I usually get away with spraying very little, but this time it did not work.

Aphids and mites

These seem to be well-controlled by natural insect enemies. I do not over-spray, but use an integrated (or intelligent) pest management program (I.P.M) and I do not have a problem with these pests. About one-third of the aphid species (beneficial ones) eat up the "bad" aphids so overspraying (such as weekly with pyrethrum) can destroy the balance. Some orchardists in large orchard areas in eastern Washington often are troubled by these two insects. Here, we are not.

Pacific flat headed borer (*Crysobothris femorata*)

About five years ago I planted a number of grafted trees, eleven I believe, on M.9 rootstock in one area and expected to transplant them the next spring to a more permanent home. For a number of reasons I was gone much of August and in a cavalier manner had not mulched these trees very thoroughly to help prevent them from becoming too dry. In late August my usual periodic examination of each tree took place. Six of these trees looked sick and there was a brown area around the graft union, which was about 1 inch above ground. I dug around in these dead tissue areas and found a light colored larva with a dark big flat head (larger than the body). All the attacks were on the south to southwest side of the small one-half inch trunk.

Pacific flat headed borers seem to attack very young apple trees, as a rule those that are under about four years of age, which still have small root structure and become drought stricken quickly. If undetected, this larva (worm) eats all the cambium area around the tree causing the tree to die. The larvae stay well inside the tree and emerge the next spring to form a small bronze to greenish-black beetle, which later deposits eggs and the cycle continues. It also attacks many stressed ornamental trees.

In 1989 and 1990 a number of horticultural articles mentioned the work of a USDA researcher who found that severely drought-stricken trees, especially young ones, gave off a sound due to broken capillaries in the tree. It seems that these sounds are at a level humans cannot hear but that this beetle can. The beetle then attacks the crying tree. One obvious solution appears to be biweekly watering and using two inches of mulch such as sawdust, barkdust or straw. These mulches prevent water evaporation and keep the tree in a more healthy mode. Some fertilizer would probably also be helpful.

The USDA is now going to go one step further and see if there is also an odor given off by the stricken tree. It may be that both are attractants to the beetle. We should know in a few years.

Another control for pacific flat headed borer is to paint the tree from about two feet above ground down to ground level with white outdoor latex paint. This reduces the warmth of the sunlight. This practice has been control enough for me if the tree is also watered regularly.

The shothole borer also attacks weakened trees and probably is controlled in a similar manner.

The aforementioned six trees had been completely girdled in the cambium area and died. The others survived and are slowly recovering while producing fruit. See the picture of the damaged area of a five-year-old producing tree that is slowly healing over (MARK rootstock).

Cold country apple culture management

These tips are for USDA zones 3 to 5, but especially for USDA zones 3 and 4.

- A conifer windbreak should be on the north and west side (or the wind side) for home orchardists or small commercial orchards. Wind can create great dehydration. Cold alone is often not critical if the wind is slowed or stopped.

- Trunks and lower limbs should be painted at planting with white latex outdoor paint to help prevent bark-splitting from winter sun-scald caused by a quick change of temperature between day and night. Repaint as the tree grows.

- Some type of supplemental water should be available in times of drought. At the time of tree dormancy and of freezing temperatures the ground should be damp, as dry roots can freeze to very low temperatures and kill the trees.

- Ensure that nitrogen levels are down to 1.8 to 2.0 percent *in the leaf analysis* by early August. High nitrogen levels markedly decrease cold hardiness, especially if nitrogen is applied after April or May. If you have rows, grow sod between the rows to about three feet from the trees. Grass helps keep nitrogen levels down.

- Do not let trees heavily over-crop, as this weakens them and reduces cold-hardiness. Three or four inches of mulch also aids survival of tree roots.

- Do not prune during cold weather, as pruning wounds undergo changes that make the pruned area much less hardy. Delayed dormancy (late March or April) is the time to prune (but don't forget summer pruning).

- High soil pH (7.5 to 9) may create problems with certain rootstocks in parts of North Dakota, eastern Montana, the prairie areas of Manitoba, Saskatchewan, and in low rainfall areas such as our southwestern states.

Columbia is reputed to be one of the better rootstocks for soils with a pH above 7.5, but consult your nearest research station if in doubt.

Soils

"Apple trees will do well in soils with a pH of 4.5 to 9." Oftentimes I have heard a scientist answer a question with the above answer and, as far as the *tree* is concerned, it is generally true; but that wasn't the question. The question was what soil pH is optimum for the production of the best *apples*: good cell structure, fine flavor, optimum keeping, etc.? From my work and soil tests in small commercial orchards, it appears that a pH of 6.3 to 7.2 produces fine apples. In my orchard, I try to keep it between 6.4 and 6.8, which I believe is optimum for most cultivars in this book.

Two types of soils *not* to use if you have any choice at all are almost pure sand, and the blue or orange heavy clay that won't come off the shovel, which creates root rots.

If your home lot has a very sandy soil and you only want a few trees, make a two- to three-foot berm of viable soil and hold it in place with, for example, two layers of railroad ties. I have done this in 85 percent of my small orchard because of heavy clay just below ground level in parts of it.

In very heavy clay that remains water-logged and which usually cannot be corrected by tiling, the railroad ties are again the answer. A commercial orchard on such heavy clay is almost surely not feasible.

In my work I have every commercial orchardist, prior to planting, use a post-hole hand digger and dig down about three feet to see what is there. Then they fill the hole with water; if it does not drain in a few hours, tiling or a shift to different soil may be in order. Rootstock selection could be important too in this case.

Except for the two extreme soils mentioned above, I can't get very excited about soils. Apple trees are amazingly adaptable to many types of sandy loam, loam, and clay loam as long as there is sufficient drainage. Here in our valley, soils tend to be heavy and should probably be tiled; however, we use that post-hole digger test before starting. If your soil is short of certain minerals, they can be added as needed after soil and leaf analysis, often by foliar feeding (applying minerals with a spray.)

In the home orchard, mulching is a must. Here we use lots of bark dust and sawdust; they keep the soil cooler and dramatically slow drying out. Less watering is thus required. Many other items also make good mulch, and straw is one favorite. A little nitrogen helps them all to break down.

If one has 12 to 18 inches of tree growth from the third year on while the tree bears apples every year, the trees probably need no nitrogen for a year or more.

Part VI

Small Commercial Orchards
(Direct Market)

I am sure that there will be many more of you wanting to grow apples at your homes than those who want a small commercial orchard; however, there are factors very pertinent to successful production from small commercial ventures that differ from home production. Perhaps I can alert you to the most important differences I have encountered. I cannot alert you to all potential problems for all areas of this diverse continent, but some points below are applicable to most areas.

1. Family considerations, with two examples:

To have a successful small direct market orchard there has to be agreement within the family about work demands, especially the marketing of fruit, or considerable family tension can occur. Most orchards of this type are started while the owner is working a full-time job, so the time is limited for him or her.

a. In the most successful two and one-half acre orchard I have ever seen, the father worked at a full-time job the first 17 years. Even with the help of his wife and three growing children, they elected to keep the orchard small and manageable. They owned more acreage but, even after retiring, the owners felt that two and one-half acres of apples were enough. This is a beautifully monitored, trellised 3- to 4-wire orchard on M.9 and M.26 rootstocks. At an average of about 35 to 45 cents a pound, this orchard's yearly gross, before costs, is about $28,000 to $32,000. For them, this is an adjunct to retirement income and is not meant to replace it. The costs of a small cold storage, electricity, some help at picking time, sprays, etc., come out of that gross.

Now that the children are grown and their help is minimal to non-existent, this orchard is all the husband and wife care to work at. More help from their neighbors is now used. Certain neighbors like having part-time income for three to four weeks of the year, which they mainly earn doing picking, cold storage boxing and some pruning. The new pneumatic and hydraulic pruning tools are time-savers and a great aid for some growers, but could be very dangerous around children or inexperienced helpers. Good old fashioned "Armstrong" (strong arm) pruning is still very effective, and perhaps more desirable in a small orchard for this reason. With minimum nitrogen use, one can usually prune half the orchard one year and the other half the next, especially the first four or five years after planting.

Most of the fruit is sold on the property, and some is sold to privately-owned grocery stores downtown, two miles away. These store owners are friends and loyally buy fruit year after year, although not in large quantities. With the help of cold storage for the long-keeping cultivars, the apples are usually kept until Christmas through the end of January, when they are sold out. They primarily use eight different cultivars, and their quality is outstanding.

Brock and Jonagold are their two most popular cultivars; others are Golden Delicious, Melrose, Idared, Braeburn, Mutsu and Fuji. The first two must be sold soon after picking, but the last six are long keepers, with Idared having fine storage properties and an improvement in flavor after a month of storage. In most years this orchard averages about 1,200 42-pound boxes per acre.

b. A family with a different approach has about 5 acres of apples, as well as growing some pears, Asian pears, pumpkins, squash and other produce. The orchard was built up about one and one-half acres each year over a period of four years. The family also has 57 acres for alfalfa hay, pumpkins, vegetables, etc., along with all the machinery and storage. They run a handsome and fair-sized country store from berry-time in June until at least the end of December. The store buys some products to sell such as honey and berries, and they make and sell a special apple butter at a fine profit. They also sell hay and straw at the store. In October, the owner invites about 2,000 grade-school and high-school students and teachers to tour his trellised apple orchard and pick out their own pumpkin for a reasonable cost; wagon rides, pulled by 2,000-pound Percherons, are a part of the event.

Except for their own store, their main outlets for apples, pears, beans, sweet corn, squash, and cider are at three Saturday markets. These markets are open each Saturday morning from summer to fall. One is in downtown Portland, Oregon, and the other two are in outlying towns. Taste tests of their apple juice are offered before buying, usually resulting in instant purchases. Taste tests are given for all apples, and all cultivars are currently priced at 60 cents a pound, meaning better quality at less than store prices. The wife takes charge of one market with one helper, and the husband takes the others with helpers. These growers also sell some apples to stores, but cannot count on consistent sales from them. They occasionally take some of their best apples to the only packer-grower sales company in the Willamette Valley; this company sells high-grade new cultivars to markets from Los Angeles to Seattle. Many of their less than perfect apples are now going to sweet apple cider.

There are many ways and places to sell good quality apples, and it is very satisfying to observe how the successful direct market owners find these sales outlets. Probably only the friendly and optimistic should try it. Don't overlook the women in the family, as often it is the wife, sister or daughter who is best at the sales. Time is critical. Who in the family has the time?

2. The cost of an orchard

Few people would suggest you start a small commercial orchard by planting on land you just purchased, then waiting three years before income. The following calculations do not take into account the price of the land itself, which in our area is usually about $3,200 to $4,200 per acre.

The Cost of an Orchard

Trees (M.26 size root) for 1 acre = $2,900 average

Water, poles, wire,
water pump, water
lines, etc., for 1 acre = $3,000 average

Totaling about $5,900 per acre,
                                         without land costs.

Can you do it for less? Yes. Scrounge! Try to use land previously purchased or that you can pay for from an income not related to the orchard. Do most or all of the labor yourself except perhaps the placing of poles. *Start small*, on one acre or less. If you trellis your trees, scrounge for wire at agricultural sales. (I do suggest trellising for many climates and soils; it pays in the long run and is necessary for the more dwarfing rootstocks.)

3. The cultivars you use

Use the best cultivars on the compatible rootstock for your climate and soil! Here is an area where serious mistakes can really cost the owner. MM.111 and MM.106 rootstocks should not be called semi-dwarfs; these two MM rootstocks should be called semi-standard in size. With large vigorous cultivars grafted to them such as Gravenstein, Melrose, Mutsu, Northern Spy and others, you will be able to harvest them only with ladders in a few years. Even worse is to plant the above 8 feet apart. Please check my planting charts. Unless planted on some very weak, rather sandy soils, I do not suggest the two MM rootstock above except for use with spur strains, since multiple spurring strains are about 25 percent dwarfing. Alternatively, use MM.111 with a dwarfing interstem.

If you want a free-standing tree, then M.7A would be your best choice. MM.111 with an interstem is also a fine choice if you can find them, but they need some support for at least a few years.

4. The sales price of apples

Currently, it appears the year-round average price of most apples in stores is about 79 cents a pound and, at times, 99 cents or more. One of the poorest decisions a grower can make is to sell high grade ripe fruit for 19 to 29 cents a pound. This person probably does not keep records or, if he does, only figures in part of his costs. There are the costs just mentioned, as well as pruning, thinning of fruit, sprays, some machinery, taxes, and probably some labor.

Why should a store gross 40 cents to 60 cents a pound simply for supplying a small cold storage and some labor cost, while a grower who takes all the risks may only net 8 to 10 cents a pound, even at 1,000 42-pound boxes an acre (42,000 pounds)? From the fifth to sixth year on, an acre of apples should gross $10,000 to $15,000 an acre. Costs come out of that. The investment cost of starting the orchard is substantial. I recommend starting small and making the orchard pay for enlargement rather than sinking a lot of cash money in each year, even if you do have it.

The above statements are based on a simple premise: this is not a short-range home hobby or direct market get-rich-quick scheme. You must pay your dues in practice and learning. To learn how to handle the trees in your climate, find out as much as you can locally before planting. Assuming good quality, selling your apples at break-even prices cheapens your whole operation and leads the buyer to believe there is very little cost to producing fine apples.

If you can control the above four points, you will probably have a successful direct market orchard.

One very experienced grower in Washington believes that with tight, double-row trellised Granny Smith on very small rootstock and 2,000 or more trees to the acre, he could gross about $30,000 to $35,000 per acre before costs. Of course, everything must be done correctly—rootstock, cultivar, pruning, etc., in order for this to happen. A novice should start with perhaps a maximum of 600 trees per acre with a small planting and learn from that.

Home orchardists, interested in only a few trees, would have quite different views from someone running a commercial venture, however small.

"Niche" apple selling

For the small commercial grower, it seems one should not try to compete too strongly with large acreage cultivars. Here in the west, Red Delicious seldom brings top prices anymore except in "extra fancy" grade. I think there are simply too many of these apples at present. So why would a grower here in the Willamette Valley plant them? We have been looking for those cultivars that do well in our cooler, long season. Here, and in northwest Washington, Jonagold is one of our best "niche" apples, but it must be sold at once after picking, with as short a storage as possible. I do not recommend shipping Jonagold overseas. It is not durable enough, even in cold storage. It is too risky and, besides, our west coast markets could use many more hundreds of thousands of bushels. Jonagold is not a hot weather apple, even though some eastern Washington growers keep struggling with it.

Braeburn also does very well here and is our great-tasting late keeper. It has been a bitterpit, internal browning disaster in the heat of southern California, but I think some of that stems from inexperienced management. Growers there like Gala, Fuji and Granny Smith as they are more manageable. Some of eastern Washington does well with Braeburn, but they seem committed to Fuji, an apple the packers and retail stores can't seem to destroy before it reaches the consumer, still crisp, juicy and sweet. Fuji, then, is the apple for them, and why not?

We have other candidates for "niche" selling, but it will take time to sort it all out. Goldrush, a late, fine-tasting, long-keeping, scab-immune apple, looks like a candidate.

I am testing Honeycrisp, a patented Minnesota newcomer. It is big, red, crisp, tasty and keeps well. It is not just a north-country apple, but the fact that it is makes it even more versatile. Cox Orange and Spitzenberg also have "niche" markets in three of our direct market orchards, with no retail store competition.

There are others, but they must be adapted to your climate and, when you find them, they must grow enough for your marketing program.

Part VII

What is my "Favorite Apple"?

When I am asked about my "favorite" apple, it is seldom clear whether I should answer with my favorite *all-purpose* apple or my favorite *tasting* apple. Although some hair splitting is involved here, there can be a discernable difference between favorite all-purpose and favorite tasting apples.

Years ago I realized that two types of people would ask me the above questions. If they are an old friend or acquaintance, they often ask just to see if I have changed my mind since foolishly stating about 20 years ago that Ashmead was my favorite tasting apple. It is still one of my favorite tasting apples, but it could not be my favorite all-purpose apple because of its small size for baking and, in cooler summers, it is a rather poor, some say ugly, color, with skin often the texture of brown sandpaper. Also, it is not quite the keeper that a Newtown or Melrose is, but it does keep well.

Most apple connoisseurs won't give a direct answer to a question such as "What is your favorite apple?" and often go to great lengths to avoid answering. One very knowledgeable man who has grown over 300 cultivars in the last 40 years has a unique, straight-faced way of laying down a 40-year smoke screen by stating: "Well, let's see. When I was seven and a half years old I ate a ripe Wealthy in September and, up to that time, it was the best apple I ever tasted." He then takes you through 5- to 10-year increments up to the present with a number of cultivars, and includes which time of the month they were picked. And there is always the one which tastes like licorice, and he likes licorice. Now his listener does not know this man probably really does not know which he likes the best because he has never had to make this decision.

Now we get to the second type who asks this kind of question. Usually they are under 45 years of age, and I know they probably can't name six apple cultivars and have no idea why the stores would not carry the very best tasting apples. In most cases this second type comes to fruit show meetings to learn and, because of limited knowledge, they don't know we are not dealing with an exact science. We are involved in highly emotional and very personal choices.

After explaining to the questioner that in western Oregon and western Washington alone, 226 cultivars of apples have appeared at fruit shows over the last 16 years and that two people seldom have the same answer to the "favorite apple" question, he or she now begins to think!

Robert Nitschke, the ultimate connoisseur who founded Southmeadow Fruit Gardens, is the only authority I know in this country who has published a favorite list of under 20 cultivars. Mr. Nitschke states that in a moment of weakness some years ago, he consented to list 12 for the *Detroit News*. Recently he has dropped three of those and upgraded it to his 13 favorites, which he recently published in Southmeadow's yearly price bulletin. He admits he still grows over 100 cultivars, and at least 40 are favorites. He said listing only a favorite 13 was akin to abandoning someone in his family. On the other hand, an Englishman, Dr. Potter, did pick his five favorites out of 2,500 cultivars.

I am going to some length to explain the difficulty of answering these questions. My selection will set a record for brevity and, by doing so, will give me a few sleepless nights and perhaps from now on a somewhat troubled life.

In the last 22 years I have planted a number of trees of each of the following cultivars: Elstar, Spitzenberg, Braeburn, and Newtown. This is preparation for the near future when my growing space will be taken up with more flower gardens and only a small area will be needed for a test plot of one tree for each cultivar, thus forcing some choices upon me.

I like to point out that my favorites are all somewhat sharp, rich, zippy, strong, and high flavored. Flavor is almost impossible to describe and nothing takes the place of taste tests.

For the second type of questioner, I offer this list of four favorites (all have color pictures and are listed alphabetically under the apple descriptions.)

The first is a tart mid-season apple that mellows in storage. The other three are late apples of great versatility for fresh dessert, sauce, pie, baking and sweet cider. I believe a favorite apple should exhibit multipurpose usage and should keep well.

Four Favorite Apples

1. **Elstar.** For sauce and pie. Some can be picked by August 20 to September 5, and rival the much more difficult to grow Gravenstein. It is picked later in cooler areas, from September 5 to 15. It is one of the finest dessert, out-of-hand, eating apples of all time. A new type and a fair keeper. Demands cool nights pre-harvest.

2. **Spitzenberg**. One has to grow this one in a long season to appreciate this "Very good to best" apple, grown for 250 years or more. It is a good keeper. Seldom grown commercially anymore because of tree disease susceptibility and strong climate preference.

3. **Braeburn**. A long keeper of tremendous flavor, and I think it is the best of the newer late apples.

4. **Newtown**. When picked ripe, after pink and purple stripes are visible, it keeps for months, even under garage conditions. Rated the highest of all, "Best." Newtown has exhibited some self-fertility.

All these cultivars bloom with good overlap in my climate, which is accidental as I would have selected them even without bloom overlap. I will probably never trim down my orchard to these four cultivars, but I could live with only these if I had to. So, now I am committed, with two very old cultivars and two newer types as my current favorites.

Appendices

Nursery Addresses

In each apple description, one or more of these nurseries is listed as a source of trees and/or rootstock. There are no doubt others, as there are hundreds of nurseries in North America. Such additional listings are beyond the scope of this book.

Adams County Nursery
P.O. Box 108
Aspers, PA 17304
(717) 677-8105

> Many apple cultivars, especially for the eastern U.S., such as Ginger Gold and Brock.

Bailey Nursery, Inc.
1325 Bailey Road
St. Paul, MN 55119
(612) 459-9744

> Very good source for cold-hardy apples on cold-hardy rootstock, as well as Malling rootstock. Virus indexed for Canadian shipment. Old, established nursery.

Bear Creek Nursery
P.O. Box 411
Northport, WA 99157

> Numerous apple cultivars, many on cold-hardy rootstock. Catalog $1.00, refundable with order.

Brandt's Fruit Trees, Inc.
P.O. Box 10
Parker, WA 98939
(509) 877-3193

> Third generation apple growers and now a nursery. Has U.S. rights on two new, very late fruit apple cultivars from Australia, Pink Lady and Sundowner.

Burchell Nursery, Inc.
4201 McHenry Avenue
Modesto, CA 95356
(209) 529-5685

> Sells Swiss Gourmet (Arlet) and Rubinette apples.

C & O Nursery
P.O. Box 116
Wenatchee, WA 98801
(509) 662-7164

> Old commercial nursery selling many apple cultivars and crab apples. Free catalog.

Carlton Plants
P.O. Box 398
Dayton, OR 97114
(503) 868-7971

> Hundred-year-old nursery. Sells apple rootstock. Also sells grafted and budded apple trees on contract. Write or phone for information.

Cloud Mt. Farm & Nursery
P.O. Box 116
Everson, WA 98247
(206) 966-5859

> Sells apple trees and some old English cider apples, plus other fruits. Catalog $1.00.

Columbia Basin Nursery
P.O. Box 48
Quincy, WA 98848
(800) 333-8589

> Sells many apple cultivars including Swiss Gourmet (Arlet) and Rubinette.

Dave Wilson Nursery
19701 Lake Road
Hickman, CA 95323
(800) 654-5854

> A very reliable 57-year-old nursery which features the apples Pink Lady, Sundowner, etc. Write for catalog.

Fruit Testing Association
  Nursery, Inc.
P.O. Box 462
Geneva, NY 14456
(315) 787-2205

> Started in 1918 and featuring new apple cultivars from Geneva Research Station, plus 8 antiques.

Greenmantle Nursery
3010 Ettersburg Road
Garberville, CA 95440
(707) 986-7504

> Best source for the "Etter apples." Fine catalog for $2.00. Owned by Ram and Marissa Fishman. Ram was a president of North American Fruit Explorers (NAFEX).

Lawsons Nursery
Route 1, Box 473
Ball Ground, GA 30107

> Apple trees for the south and southeast U.S. Some cultivars are quite old and hard to find, such as Yates.

Lawyer Nursery, Inc.
Plains, MT 59859
(406) 826-3881

> An extensive source of cold-hardy rootstock such as Borowinka, Antonovka, Columbia, etc. Also sells grafted trees. Virus indexed for foreign shipment. Write for free plant list.

Living Tree Center
P.O. Box 1002
Berkeley, CA 94709
(510) 420-1440

> Many old apple cultivars. Has Reinette Simerenko on MM.111. Valuable catalog, $4.00.

Meadow Lake Nursery Co.
P.O. Box 1302
McMinnville, OR 97128
(503) 852-7525

> Grows many different apple rootstocks including some of the new Geneva series. Virus indexed for foreign shipment. Free catalog.

Northwoods Retail Nursery
27635 South Ogelsby Road
Canby, OR 97013
(503) 266-5432

> Excellent 60-page catalog of "unique fruits, nuts, and ornamentals for the home gardener." Free catalog.

Raintree Nursery
391 Butts Road
Morton, WA 98356
(206) 496-6400

> Seventy-page informative catalog of fruits, nuts and berries for the Pacific Northwest, especially west of the Cascade Mountain Range. Free catalog.

Roaring Brook Nurseries
Route 1, Box 186
Monmouth, ME 04259
(207) 375-4884

> Small nursery in northern New England, and one of the few that sells Brock, an apple for USDA zone 4 in the east, and for cooler zones 7 and 8 in the west.

Rocky Meadow Orchard and
  Nursery
360 Rocky Meadow N.W.
New Salisbury, IN 47161
(812) 347-2213

> Small nursery managed by Ed Fackler, one of the Midwest's real apple experts. Sells rootstock and hard-to-find Freyberg, Brock, etc. Catalog $2.00.

Rutgers Fruit Research and Development Center
c/o Joseph Goffreda
R.R. #539
Cream Ridge, NJ 08514

> Source of Suncrisp.

Sonoma Antique Apple Nursery
4395 Westside Road
Healdsburg, CA 95448
(707) 433-6420

> Many old apple cultivars and a few modern ones. Other fruits too. Also old English cider apples. Catalog $2.00, refundable with order. Catalog explains which are low-chill cultivars.

Southmeadow Fruit Gardens
c/o Grootendorst Nursery
Lakeside, MI 49116
(616) 469-2865

> More than 200 grafted apple cultivars originating from all over the world. Fine source of Malling rootstock, especially M.9. Also has interstems. Apple plant list free.

St. Lawrence Nurseries
Route 2
Potsdam, NY 13676
(315) 265-6739

> A zone 3 north-country nursery using organic methods. Most of their 100 apple cultivars are on cold-hardy rootstock. Free catalog.

Stark Bros. Nurseries
P.O. Box 10
Louisiana, MO 63353
1-800-325-0611
(314) 754-5511

> Our oldest continuous nursery, dating from 1816. Source of many apple cultivars. Now again growing the original Hawkeye delicious in limited quantity. Free catalog.

Van Well Nursery
P.O. Box 1339
Wenatchee, WA 98807
(509) 663-8189

> Old family nursery with cultivars Ginger Gold, Lustre Elstar, etc. Free color catalog.

Canadian nurseries

(For others, consult your province university.)

Corn Hill Nursery
R.R. #5
Petitcodiac, New Brunswick
EOA 2HO
(506) 756-3635

> One of the best Canadian sources of apples on cold-hardy rootstock such as Beautiful Arcade, Borowinka, Ottowa 3, etc. Fine catalog.

Tsolum River Fruit Trees
Box 68
Merville, British Columbia
VOR 2MD
(604) 537-8004

> Organic grower. Over 150 apple and crab apple cultivars. Worldwide selection. Not virus indexed. Ships only within Canada. Catalog $3.50.

References

Apples of New York (Volumes I and II)
Authors: S.A. Beach, N.O. Booth, O.M. Taylor.

> Report of the New York Agricultural Experiment Station for the year 1903. Printed by J. B. Lyon Company, 1905. Book copies are now rare and expensive.

Compact Fruit Tree (journal, published yearly)
The International Dwarf Fruit Tree Association
Room 301-B, Horticulture
Michigan State University
East Lansing, MI 48824-1112

> The journal comes with membership. Currently $50 yearly in the U.S. for individuals; more in Canada.

Granny Smith Apple
Extension Bulletin #0814
Author: James Ballard
Washington State University
Pullman, WA

North American Apples
Many authors contributed
Michigan State University Press, 1970
East Lansing, MI 48824-1112

Western Fruit Book
Author: E.J. Hooper
S.C. Griggs and Company
Cincinnati, OH

> Published in 1857. Book copies are now very rare.

Agriculture Research Services
U.S. Department of Agriculture
Washington, DC 20402

> Statistics on apple production and other agricultural products, etc.

Bibliography

Books and annual periodicals

Training and Pruning Apple and Pear Trees
Forsley, Elving and Stebbins, 1992

American Society for Horticultural Science
113 South West Street, Suite 400
Alexandria, VA 22314-2824

> New, in-depth work on pruning; perhaps the best one yet. A difficult subject. Costs around $30.

Compact Fruit Tree (annual)
International Dwarf Fruit Tree Association
14 South Main Street
Middleburg, PA 17842

> Primarily for serious commercial growers. This yearly publication is very detailed, international in scope, and may confuse home orchardists with only a few trees. I personally find it worth the $50.00 annual fee.

Pomona Book Exchange
Highway 52
Rockton, Ontario, CANADA
(519) 621-8897

> Over 1,000 books, periodicals, etc., for sale in the fields of horticulture, agriculture, botany and related subjects. A unique North American service with literature from all over the world, some rare. Write for catalog (free).

Quarterly or monthly fruit publications

Pomona
North American Fruit Explorers (NAFEX)
Attn: Jill Vorbeck
Route 1, Box 94
Chapin, IL 62628
(217) 245-7589

> Over 3,000 amateur members, some very dedicated and knowledgeable, plus interested scientists. All contribute to a very helpful quarterly booklet. Current yearly membership fee is $11.00.

Pome News
Home Orchard Society
P.O. Box 230192
Tigard, OR 97281-0192

> A 20-year-old organization. Members are primarily amateurs, with an interesting and worthwhile quarterly publication. They have a yearly fruit show and a March meeting in the Portland, Oregon, area for purchase of small amounts of rootstocks, mostly apple and pear. Yearly membership is $10.00.

Good Fruit Grower
1005 Tieton Drive
Yakima, WA 98902
(509) 575-2315

> Semi-monthly, to sometimes monthly. A large-sized magazine primarily for western U.S. and commercial growers, with much useful information, some worldwide in scope. Yearly fee is $30.00.

Great Lakes Fruit Growers News
343 South Union Street
Sparta, MI 49345
(616) 887-8615

> A large monthly news bulletin publication primarily for the upper Midwest. Informative, with much local news. $18.00 for 3 years.

A unique service

Apple Source
Route 1, Box 94
Chapin, IL 62628
(217) 245-7589

> A small booklet describing a most unusual service: they enable you to taste apples, sent to you by mail, with information on how to order. Thus, you can taste an apple new to you before ordering any trees. They have dozens of cultivars from which to select. This is not a nursery, and they do not sell trees—just apples. The booklet is free. What a fine idea!

Glossary

Bitterpit: A physiological condition due to a calcium shortage in the apple, not necessarily in the tree or soil. (See in-depth explanation under "Culture and Management.") Apples exhibit small, round, usually dark-colored dead depressions, especially on the calyx end and sometimes all through the apple flesh.

"Cork Spot" of York is thought to be caused by mineral shortages, probably boron and calcium. Genetics, too much nitrogen, water shortage, extreme pre-harvest temperature, young trees, all contribute to bitterpit. Cooler climates, like mine, generally can grow more different apple cultivars of high color and quality than a hot valley like some in California. Heat and dehydration indirectly contribute to bitterpit.

Budding: See also "scion wood." Budding is another form of grafting, which is done in the late summer with the current summer's bud growth. The bud should simply heal, but not grow a tender branch that might freeze. Budding to a certain rootstock is usually done in late August or early September in our climate, late enough so a branch won't grow from the bud until next spring. One bud from a certain cultivar is inserted in the bark a few inches above ground on a known rootstock. The following spring, the rootstock is then cut off just above this bud. That bud then will grow into a tree genetically like the cultivar from which you selected it—all from one small bud. It seems like a miracle! In my opinion, it *is* a miracle.

Clonal rootstock: Rootstock genetically alike. Rootstock is usually grown by layering branches under soil or sawdust so they will root down, or by planting cuttings from a specific cultivar. This allows us to reproduce a rootstock identical to the mother plant. Planting a seed will not create a clone, as apple seeds do not breed true.

Common cold storage: Where apples, and some other fruits such as pears, are stored at close to 30° or 32°F to slow ripening; there is enough sugar in apples to prevent freezing at this temperature. At 40°F, ripening is greatly speeded up. Ethylene gas is given off during ripening in storage and could harm cold storage bareroot trees. It is best to store each fruit alone. In the home refrigerator, one can lengthen storage time by putting apples in plastic bags with small holes in the bag. This keeps moisture in and delays dehydration and wrinkling, especially in Golden Delicious and Gala, which can wrinkle badly at room temperature.

Controlled atmosphere (C.A.) storage: A marvelous invention for putting the apple to sleep. Ripening is almost stopped for months in many cultivars. The C.A. building must be airtight and well insulated. It is monitored by controls on the outside wall with graphs, charts, etc. Oxygen and carbon dioxide are decreased to rather small amounts. Most cultivars are kept at 32°F.

Cultivar (cultivated variety): We may be doing some hair-splitting here, but the accepted botanical way to describe a grafted tree from the original mother tree is a "cultivated variety." The original mother tree is a new variety. All reproduction later is a cultivated variety—a cultivar.

Open Pollinized: Many planted apple seeds (female is seed) from a known cultivar are pollenized by unknown apple pollen that an insect usually brings in and deposits on the flower.

An example of this would be Cox's Orange, which came from a seed of open-pollinized Ribston.

The above term should be used where the seed parent or plant is known; otherwise, parentage should be listed as "unknown" when neither the female nor male parentage is certain. Stayman is another example where the seed parent (Winesap) was known but the pollen parent was unknown (open field pollenized).

In many apples, such as McIntosh and Red Delicious, neither parent is known—suspected maybe, but not known.

Pyriform: This term was used to describe the tree of Kandil Sinap. If this tree's branches are not tied down to horizontal, they all grow straight up after coming out from the trunk a few inches. The tree looks like the apple: long and narrow. It then resembles the shape of a Lombardy poplar. Northern Spy and Newtown also tend to this type of growth unless the branches are tied or weighted down.

Scion wood (for grafting and budding): See also "budding." A bud-stick of two or three buds (the scion) taken in January or February from the new growth of the last summer, to be used for grafting to a certain rootstock in March or April. (I store mine in the refrigerator in a damp piece of paper wrapped in plastic.) This is one way we perpetuate a known cultivar, because apples now have great genetic diversity and a seed planted is unlikely to produce an edible apple. This is called "cloning" the variety.

It is not a purpose of this book to teach propagation of rootstock or the mechanics of grafting. Most state extension offices have excellent free or low-cost booklets or pamphlets for the study of grafting, and sometimes also offer courses. Apples are probably the easiest fruit to graft and I highly recommend trying it.

Sunscald (winter sunscald): Sun shining on the bark of a tree in winter daytime and warming those areas, then nighttime temperatures dropping many degrees, often 30° or 40°F or more. That sunstruck area can lose hardiness and may suffer from intercellular freezing and death. Then that part of the bark can exhibit a dead area the next summer. Winter sunscald endangered the original Delicious tree one cold Iowa winter (see description of Red Delicious.) White outdoor latex paint on the trunk and larger branches helps protect the tree by light reflection in all seasons.

This winter injury usually occurs on the south-facing tree trunk or branches.